New Frontiers in Urban Analysis

In Honor of Atsuyuki Okabe

New Frontiers in Urban Analysis

In Honor of Atsuyuki Okabe

Edited by
Yasushi Asami
Yukio Sadahiro
Toru Ishikawa

CRC Press
Taylor & Francis Group
Boca Raton London New York

CRC Press is an imprint of the
Taylor & Francis Group, an **informa** business

CRC Press
Taylor & Francis Group
6000 Broken Sound Parkway NW, Suite 300
Boca Raton, FL 33487-2742

First issued in paperback 2019

ISBN-13: 978-1-4398-0252-6 (hbk)
ISBN-13: 978-0-367-38548-4 (pbk)

Library of Congress Cataloging-in-Publication Data

New frontiers in urban analysis : in honor of Atsuyuki Okabe / Yasushi Asami, Yukio Sadahiro, Toru Ishikawa.
p. cm.
Includes bibliographical references and index.
ISBN 978-1-4398-0252-6 (hbk. : alk. paper)
1. City planning. 2. Land use, Urban. 3. Spatial analysis (Statistics) I. Asami, Yasushi, 1960- II. Sadahiro, Yukio, 1966- III. Ishikawa, Toru, 1971- IV. Okabe, Atsuyuki, 1945- V. Title.

HT166.N437 2009
307.1'216--dc22 2009015404

Visit the Taylor & Francis Web site at
http://www.taylorandfrancis.com

and the CRC Press Web site at
http://www.crcpress.com

Contents

SECTION II Urban Analysis in the Social Sciences

Toru Ishikawa

SECTION III Spatial Analysis

Yukio Sadahiro

Preface

Yasushi Asami, Yukio Sadahiro, and Toru Ishikawa

This book is a collection of chapters about urban analysis, the field of study analyzing phenomena appearing in urban space. One of the originators in this field in Japan is Atsuyuki Okabe, who was also a founder of the Center for Spatial Information Science at the University of Tokyo. The authors of the chapters of this book have benefited from his work, learning basic principles of urban analysis and related academic areas from him. Thanks to Atsuyuki Okabe a number of academic contributions have been made in Japan, but despite the high quality of these contributions, many of them are not accessible to foreign researchers and students because they have been published in Japanese. Atsuyuki Okabe's successors planned this publication in his honor at the time of his retirement from the University of Tokyo in March 2009.

The book has three main sections. The first section covers urban analysis for planning theories. Actually, urban analysis in Japan emerged from the city planning field. In the genesis of urban analysis, most of the analytical methods were borrowed from other disciplines, such as economics, quantitative geography, operations research, and spatial statistics. These methods, based on the theories of other fields, however, were not guaranteed to fit into the process of urban planning practice or to generate new insights into city planning. And indeed, a major criticism of studies in urban analysis up to the 1980s was that urban analysis did not contribute to actual city planning. It was also true that city planning in the 1960s and 1970s was not advanced enough to incorporate quantitative methods in the planning process, and moreover, quantitative spatial databases covering urban spaces or cities were not well established. At that time, quantitative studies covered only population predictions using traditional methods, such as simple trend analysis or cohort analysis, land use prediction based on the Lowry model (1964), and other simple methods.

The development of urban databases and the application of geographic information systems (GIS) to city planning changed the situation drastically. Japan became one of the leading countries in maintaining detailed spatial databases. In earlier times, however, such spatial data were mostly used to grasp the current situation at a glance: graphical presentations helped planners greatly, but sophisticated methods were rarely used in the planning process. But later on, particularly after the year 2000, when the Japanese Cabinet encouraged regulatory reform, public policies became the subject of quantitative impact analysis, with urban policies being one of them. The collection of chapters in this section is related to land use patterns, allocation of urban facilities, and urban environments, which give rise to suggestions about directions for planning.

The second section relates to urban analysis in the social sciences. Chapters in this section illustrate close relationships between urban analysis and various fields in the social sciences. Traditionally researchers in urban analysis have learned a lot from theories and methods of regional science and urban economics and have studied a variety of topics in urban space from the point of view of economics, such as bid rent, location models, housing market analysis, and residential and regional planning. The first chapter exemplifies what urban analysis researchers do in the field of urban housing science.

Another field to which researchers in urban analysis contribute and owe a lot is geography. Particularly since the quantitative (or theoretical) revolution of the late 1950s and early 1960s, researchers in various disciplines have realized the necessity of studying spatial phenomena scientifically, and urban analysis can be said to have its roots in this tradition. Among the most extensively studied topics in such spatial-science perspectives are spatial distributions and urban structure (this is particularly true with the advent of geographic information systems). The second chapter shows how researchers trained in urban engineering make contributions in this regard.

Research in urban analysis also deals with cognitive and behavioral aspects of urban space as important issues of theoretical interest and of practical value for effective urban planning. The third chapter looks at consumers' evaluations of department stores, on the basis of empirical behavioral data. The fourth chapter about environmental psychology examines how people in urban space perceive spaciousness by conducting a controlled experiment with human participants. These chapters are good examples of how urban analysis researchers study the interaction between humans and the physical environment.

The third section concerns spatial analysis. Spatial analysis emphasizes methodological aspects rather than applications, partly because it aims to analyze any phenomenon occurring in space. Applications of spatial analytical methods are not limited to urban space, but cover a wide range of academic fields, including epidemiology, ecology, sociology, oceanography, and archaeology.

Spatial analysis, in the broad sense of the term, includes exploratory spatial analysis, confirmatory spatial analysis, spatial modeling, and spatial planning. Exploratory spatial analysis aims to find interesting and important research hypotheses in order to reveal the underlying structures behind spatial phenomena. Confirmatory spatial analysis examines the validity of hypotheses obtained by exploratory analysis. Spatial modeling simulates spatial phenomena, usually described quantitatively, to predict their future conditions and evaluate the impact of spatial environments. Using the results of spatial modeling, spatial planners seek appropriate actions on spatial phenomena.

Chapters in this section propose new methods of spatial analysis and apply them to the analysis of urban phenomena. Some chapters emphasize the development of methods, while others focus on applications. This section provides a concise but sufficient outline of the field of spatial analysis. It should attract readers from disciplines outside urban analysis, such as epidemiology and archaeology, as mentioned earlier.

In short, although this book comprises a collection of chapters in urban analysis that are mostly from Japan, we believe that the general nature and high quality of the contributions will attract readers from all over the world, not only in the field of urban analysis, but also in related fields.

Future Directions of Spatial Analysis

Atsuyuki Okabe

I am honored that this volume has been compiled in commemoration of my retirement from the University of Tokyo. I express my deepest thanks to the contributors and the editors, Yasushi Asami, Yukio Sadahiro, and Toru Ishikawa. In reading each chapter in this volume, I fondly recalled the good old days when I enjoyed discussions with the contributors on or about urban analysis, one of my classes, and during the development of their theses. I am very pleased that they are now active in their own fields and developing new frontiers of urban analysis, as reflected in the title of this volume. With this as my inspiration, I personally envisage the future directions of urban analysis or, more broadly, spatial analysis, along with revisiting my past studies.

I believe there are two complementary but differing directions for spatial analysis. Analogically speaking, a high mountain has a broad foot: if the foot is narrow, the top cannot be high. Spatial analysis then becomes richer when "spatial analysis at the top" is associated with "spatial analysis at the foot." In turn, spatial analysis at the foot is spatial analysis for the nonspecialist. A few decades ago, only researchers in a very restricted field at the university level enjoyed spatial analysis. These days, spatial analysis is increasing among researchers not traditionally specialized in this area and among nonresearchers (including government officers, volunteers in nonprofit organizations (NPOs), facility managers, realtors, and developers) wishing to carry out spatial analysis in their daily business. However, spatial analysis has not always responded to these demands. To fill this gap, there is the expectation that spatial analysis at the foot will develop toolboxes with which everyone can easily perform spatial analysis. At present, such toolboxes are just not good enough. A more successful example is statistics, a field employed by a wide range of users each day simply because user-friendly toolboxes, including Excel and SPSS (Statistical Package for the Social Sciences), are readily available at reasonable prices.

A decade ago, I noted the importance of the tools for spatial analysis and came to believe that spatial analysis using geographical information systems (GIS), or what I call *GIS-based spatial analysis*, would make spatial analysis friendlier to researchers presently analyzing spatial factors using tedious and laborious manual methods. In 1997, along with Yasushi Asami and Yukio Sadahiro, I attempted to bring GIS-based spatial analysis into a new field of application: Islamic area studies. In practice, this introduction was not straightforward. Any tool is useless unless users can devote the time necessary to learn what it is, what it can do, and how to use it. In this case, collaboration between the GIS-based spatial analysts and non-GIS-based researchers in Islamic area studies was indispensable (Okabe, 2004).

Fortunately, we had good collaboration with the Islamic area studies project (led by Tsugitaka Satoh), even though at the outset of the project few researchers in the field knew much about GIS-based spatial analysis and vice versa. A few years were required for this learning stage, during which we initiated collaborative work convinced that GIS-based spatial analysis could be effective for Islamic area studies. This provided us with new ways of thinking and enabled us to detect new aspects of Islamic area studies that would not otherwise have been discovered. The crystallization of the outcome was the publication of *Islamic Area Studies with Geographical Information Systems* (Okabe, 2004), awarded the World Prize for the Book of the Year by the Islamic Republic of Iran. Encouraged by this success, I commenced a project to spread GIS-based spatial analysis to the humanities and social sciences. Through this project, application was made of GIS-based spatial analysis to history, archaeology, sociology, economics, psychology, and regional hygiene in Japanese academic communities. Eventually, the applications appeared as *GIS-Based Studies in the Humanities and Social Sciences* (Okabe, 2005).

Through these projects, I realized that while GIS-based spatial analysis helped researchers in many academic fields, it was of little help to nonresearchers, including government officers, NPO volunteers, facility managers, realtors, and developers. Traditional academic spatial analysis assumes a homogeneous plane with Euclidean distance. However, in the real world, most facilities are located along streets, and people access amenities through a street network. I realized that the managers of facilities wanted more realistic spatial analysis. To respond to this demand, in the late 1990s, Keiichi Okunuki, Shino Shiode, and I developed a toolbox (SANET) for spatial analysis of a network embedded in two- and three-dimensional space (Okabe et al., 2006). This toolbox has extended the use of spatial analysis to nonresearchers as well as to researchers not specialized in spatial analysis. This particular toolbox is in use in more than forty countries and has found many applications, including by public officers managing traffic accidents and street crime. Although toolboxes for spatial analysis have developed to a certain extent, progress is slow. One reason may be that researchers tend to not consider the development of spatial analysis at the foot to be a frontier development. I do not agree, and hope that more spatial analysts will participate in this frontier development, and that the day will come when everybody in everyday life enjoys spatial analysis.

Spatial analysis of the second direction, or spatial analysis at the top, refers to advanced spatial analysis for the specialist. I note that spatial analysis at the top will eventually apply to spatial analysis at the foot. There are many attractive peaks in the mountain range of spatial analysis. Let me follow the trail heading to one of these peaks. Spatial analysis began with *macroscale spatial analysis*, such as Christallar's city systems, where points with their attributes (e.g., locations of cities with their populations) represent cities. I very much enjoyed the topic of macroscale spatial analysis in my second doctoral thesis titled "Contributions to the Studies of City Size Distributions" (1977). The contributions in this thesis were later developed into a series of papers: Okabe (1977, 1979, 1987) and Okabe and Sadahiro (1996).

Macroscale spatial analysis next developed to *mesoscale spatial analysis*, where a set of spatial units forming a tessellation of the region represents some part of the real world, such as administrative districts, and the attributes of each spatial unit aggregate across each spatial unit (e.g., total population in each administrative district). The Voronoi diagram, which I studied with Barry Boots, Kokichi Sugihara, and Sun Nog Chiu, has been utilized for mesoscale spatial analysis in many natural as well as social sciences. In fact, more than 2,000 papers cite our book (Okabe et al., 1992, 2000). In 1992, spatial analysis using the Voronoi diagram was spatial analysis at the top, but nowadays it has become spatial analysis at the foot in the sense that it is now used as a tool in many studies in the natural and social sciences.

In recent years, spatial analysis has developed into *microscale urban analysis*, where the real world is represented by a set of many types of objects (including moving objects—consequently, *spatiotemporal analysis*) with their detailed attributes. GIS have greatly supported the development of microscale spatial analysis, in particular in data acquisition with advanced information technology (e.g., many types of sensors), data management employing huge databases, visualizations via multimedia, and communication via the Internet.

In contrast, however, spatial analysis itself, especially statistical analysis, is lagging behind. Major statistical methods for spatial analysis still assume aggregated data, and there are not as many statistical methods for analyzing individual objects in a detailed space. Observing that daily individual activities in buildings, districts, cities, and regions are performed through networks, e.g., pathways, streets, railways, waterways, and the Internet, one of the first needs was to establish statistical spatial analysis in networks embedded in two- and three-dimensional space. I began to develop this in 1995 alongside Hidehiko Yomono and Masayuki Kitamura (Okabe et al., 1995; Okabe and Kitamura, 1996), and subsequently developed a series of statistical methods: Okabe et al. (1996, 2008, 2009), Okabe and Okunuki (2001), Okunuki and Okabe (2002), Okabe and Yamada (2001), and Okabe and Satoh (2005). In addition to statistical methods, we should develop microscale spatial analysis using diverse methods. In particular, behavioral and psychological methods are increasingly important to deal with individual behavior supported by modern information devices (Ishikawa et al., 2008). At present, many theoretical subjects in microscale spatial analysis are still awaiting development.

We are now in the midst of an ongoing revolution brought about by information and communication technology (ICT). In the future ICT society, microcomputers, computer tips, tags, and geosensors will be embedded in many objects in our environment. Ken Sakamura refers to a society enriched by such a system as the *ubiquitous computing society*. We can see signs of this society today: many people are constantly using cell phones for action at any time and in any place. To realize the ubiquitous computing society, we need to develop a system in which, at any time and in any place, everybody can receive the most appropriate personalized information for action, considering his or her circumstances at that time and in that place.

To construct this system, we expect to develop new frontiers of spatial analysis, in which we analyze the circumstances of an acting body (including a person, a group of persons, a company, and in the future, a robot) and derive appropriate personalized information for action *almost in an instant*. I refer to this as *real-time spatial analysis*. This appears only as a dream today because of the requirement for a new social infrastructure. Fortunately, the establishment of WG10 (Public Fee Access) of ISO/TC211 in 2008 is a step toward realizing this infrastructure and the ubiquitous computing society is now in sight. Why not accept this challenge to develop real-time spatial analysis?

ACKNOWLEDGMENT

This preface is partly based on a 2006 e-interview with Dr. Okabe by Dr. Jungyeop Shin (Okabe, 2006) and the preface of Okabe (2004) and Okabe (2009).

REFERENCES

Ishikawa, T., Fujiwara, H., Imai, O., and Okabe, A. 2008. Way finding with a GPS-based mobile navigation system: A comparison with maps and direct experience. *Journal of Environmental Psychology* 28:74–82.

Okabe, A. 1977. Some reconsiderations of Simon's city-size distribution model. *Environment and Planning A* 9:1043–53.

Okabe, A. 1979. An expected rank-size rule: A theoretical relationship between the rank-size rule and city size distributions. *Regional Science and Urban Economics* 9:21–40.

Okabe, A. 1987. A theoretical relationship between the rank-size rule and Clark's law of urban population distribution: Duality in the rank-size rule. *Regional Science and Urban Economics* 17:307–19.

Okabe, A. ed. 2004. *Islamic area studies with geographical information systems*. London: RoutledgeCurzon.

Okabe, A. 2005. *GIS-based studies in the humanities and social sciences*. Boca Raton, FL: CRC/Taylor & Francis.

Okabe, A. 2006. Network-based spatial algorithm. In *Planning and Policy*, Korea Research Institute for Human Settlement, November, 81–97.

Okabe, A. 2009. Special issue, *Journal of Geographical Systems.*

Okabe, A., Boots, B., and Sugihara, K. 1992. *Spatial tessellations: Concepts and applications of Voronoi diagrams*. Chichester, UK: John Wiley.

Okabe, A., Boots, B., Sugihara, K., and Chiu, S. N. 2000. *Spatial tessellations: Concepts and applications of Voronoi diagrams*. 2nd ed. Chichester, UK: John Wiley.

Okabe, A., and Kitamura, M. 1996. A computational method for market area analysis on a network. *Geographical Analysis* 28:330–49.

Okabe, A., and Okunuki, K. 2001. A computational method for estimating the demand of retail stores on a street network and its implementation in GIS. *Transactions in GIS* 5:209–20.

Okabe, A., Okunuki, K., and Shiode, S. 2006. The SANET toolbox: New methods for network spatial analysis. *Transactions in GIS* 10:535–50.

Okabe, A., and Sadahiro, Y. 1996. An illusion of spatial hierarchy: Spatial hierarchy in a random configuration. *Environment and Planning A* 28:1533–52.

Okabe, A., Satoh, T., Furuta, T., Suzuki, A., and Okano, A. 2008. Generalized network Voronoi diagrams: Concepts, computational methods, and applications. *International Journal of Geographical Information Science* 22:1–30.

Okabe, A., and Satoh, T. 2005. Uniform network transformation for points pattern analysis on a non-uniform network. *Journal of Geographical Systems* 8:25–37.

Okabe, A., Satoh, T., and Sugihara, K. 2009. A kernel density estimation method for networks, its computational method, and a GIS-based tool. *International Journal of Geographical Information Science*, to appear.

Okabe, A., and Yamada, I. 2001. The K-function method on a network and its computational implementation. *Geographical Analysis* 33: 271–290.

Okabe, A., Yomono, H., and Kitamura, M. 1995. Statistical analysis of the distribution of points on a network. *Geographical Analysis* 27:152–175.

Okunuki, K., and Okabe, A. 2002. Solving the Huff-based competitive location model on a network with link-based demand. *Annals of Operations Research* 111:239–52.

About the Contributors

Atsuyuki Okabe received his PhD from the University of Pennsylvania in 1975 and the doctor of engineering degree from the University of Tokyo in 1977. Previously he held the position of associate professor at the Institute of Socio-Economic Planning, University of Tsukuba, and professor in the Department of Urban Engineering, the University of Tokyo. He is currently professor at Aoyama Gakuin University, and a member of the Science Council of Japan. He was director of the Center for Spatial Information Science from 1998 to 2005, and is now project professor at the Center. He specializes in spatial analysis and geographic information science. One of his coauthored books is *Spatial Tessellations: Concepts and Applications of Voronoi Diagrams* (2nd ed., John Wiley, Chichester, UK, 2000).

EDITORS

Yasushi Asami graduated from the University of Tokyo in 1982 and earned a PhD from the University of Pennsylvania in 1987. After becoming an assistant professor (in 1987), lecturer (in 1990), and associate professor (in 1992), he is now a professor at the Center for Spatial Information Science, the University of Tokyo, and also vice director of the center (from 2005). His research topics range from urban housing and city planning to spatial information analysis.

Yukio Sadahiro graduated from the University of Tokyo in 1989. He became an assistant professor in the Department of Urban Engineering, the University of Tokyo, in 1991. He moved to the Research Center for Advanced Science and Technology in 1995, the Center for Spatial Information Science in 1998, and then returned to the Department of Urban Engineering in 2001 as an associate professor. His research interest covers almost the whole range of spatial analysis, from exploratory spatial analysis to spatial models and decision making.

Toru Ishikawa is an associate professor at the Graduate School of Interdisciplinary Information Studies and the Center for Spatial Information Science, the University of Tokyo. He received a PhD in geography from the University of California, Santa Barbara, and was a postdoctoral and associate research scientist at the Lamont-Doherty Earth Observatory, Columbia University. His research interests are in various aspects of human spatial cognition and behavior, and he has published papers in journals such as *Cognitive Psychology*, *Cartography and Geographic Information Science*, the *Journal of Geoscience Education*, the *Journal of Environmental Psychology*, *Environment and Behavior*.

CHAPTER AUTHORS

Akio Hori graduated from Tokyo Metropolitan University in 1998 and earned a master of engineering from Tokyo Metropolitan University in 2000. After working in Hamagin Research Institute, Ltd. (1998–2005), he is now working in Mitsubishi Research Institute, Inc. (2005–present). His research topics range from quantitative analysis on credit risk of financial institutions, and data mining to location–allocation planning of regional facilities.

Fumiko Ito graduated from the University of Tokyo in 1988 and earned a doctoral degree (engineering) from the University of Tokyo in 1997. After becoming a research associate (in 1997) at the Science University of Tokyo and associate professor (in 2003) at the Graduate School of Modern Society and Culture, Niigata University, she is now an associate professor in the Department of Urban Science, Tokyo Metropolitan University (since 2006). Her research topics range from urban housing and urban structure to spatial analysis.

Shigeru Kakumoto was a senior researcher in the Hitachi Central Research Laboratory, and is now a team leader of the Information Technology Implementation for Disaster Mitigation Research Team at the Earthquake Disaster Mitigation Research Center, National Research Institute for Earth Science and Disaster Prevention. His research interests include image processing, automatic map recognition systems, and geographic information systems. His current research focuses on constructing multicolored figure input systems and four-dimensional geographic information systems (GIS) that take time into account.

Teruhisa Kamachi graduated from the University of Tokyo in 1986. After working as a researcher in NEC Corporation and also at Sony Corporation, he is now a general manager of So-net Entertainment Corporation (since 2005). His research topics range from virtual society to network entertainment.

Yoshiaki Kume graduated from the Faculty of Engineering in 1980 and the School of Engineering in 1982, University of Tokyo. He also earned a doctor of engineering in 2006 from the University of Tokyo. After being engaged as a researcher in Mitsubishi Research Institute, Inc. for 17 years, he became a professor in the Department of Urban Economics, Nasu University, whose name was changed to the Department of City Life Sciences, Utsunomiya Kyowa University. He is now a professor at the National Graduate Institute for Policy Studies.

Satoru Masuda graduated from the University of Tokyo in 1982 and earned a PhD from the University of Tokyo in 1987. After becoming a researcher at Mitsubishi Research Institute, Inc. in 1987, lecturer at Tohoku University (the College of General Education) in 1990, and associate professor (the Graduate School of Information

Sciences) in 1993, he is now a professor at the Graduate School of Economics and Management, Tohoku University (since 2000). His research fields are regional planning and risk management policy for hazard mitigation.

Akiko Murata graduated from the Science University of Tokyo in 1998 and earned a master's degree (engineering) from the Science University of Tokyo in 2000. After becoming an engineer at Nerima City (in 2000), she is now a chief engineer of the City Planning Section in the Division of City Development in the Nerima City Office (since 2008).

Kei-ichi Okunuki graduated from the University of Tokyo in 1991 and earned a PhD from the University of Tokyo in 1997. After becoming an assistant professor (in 1995) in the Department of Urban Engineering, the University of Tokyo, he is now an associate professor in the Department of Geography, Nagoya University (since 1999). His research topics range from locational optimization to spatial analysis.

Toshinori Sasaya graduated from the University of Tokyo in 2006. He earned his master's degree from the University of Tokyo in 2008. His graduate and master's theses discuss the relationship among spatial tessellations, part of which is included in the present book. He is now working in the Technology Planning Department of the Mathematical Science Section at Tokyo Gas.

Narushige Shiode is lecturer in spatial analysis and GIS at Cardiff University. He was previously a research fellow at University College London (2000–2003), then an assistant professor of geography at the State University of New York at Buffalo (2003–2008). He holds a PhD in architecture (University of London) and bachelor's and master's degrees in urban planning from the University of Tokyo. His research is focused around spatial analysis and modeling of urban spaces, with applications in urban growth dynamics, three-dimensional visualization, and GIS.

Shino Shiode is currently a visiting research fellow at the Center for Spatial Information Science, the University of Tokyo. She obtained her PhD in spatial analysis from the University of Tokyo, a master's degree in urban planning from the University of Tokyo, and a bachelor's degree in economics from Gakushuin University. Her research fields are spatial analysis and GIS with a focus on urban crime and health.

Tsutomu Suzuki graduated from the University of Tokyo in 1987 and earned a doctoral degree (engineering) from the University of Tokyo in 1995. After becoming an assistant professor (in 1996) and associate professor (in 2003) at the Institute of Policy and Planning Sciences, University of Tsukuba, he is now a professor in the Department of Risk Engineering at the Graduate School of Systems and Information Engineering, University of Tsukuba (since 2005). His research topics range from location analysis, spatial analysis, and urban structure to transportation modeling.

Hidenori Tamagawa graduated from the University of Tokyo in 1980 and earned a doctor of engineering from the University of Tokyo in 1987. After becoming an assistant professor (in 1984) and associate professor (in 1991) in the Department of Architecture, Niigata University, associate professor (in 1994) at the Center for Urban Studies, Tokyo Metropolitan University, and professor (in 1999) at the Graduate School of Urban Science, Tokyo Metropolitan University, he is now a professor in the Department of Urban System Science at the Graduate School of Urban Environmental Sciences, Tokyo Metropolitan University.

Daisuke Watanabe graduated from the University of Tsukuba in 1999 and earned a doctoral degree (engineering) from the University of Tsukuba in 2006. After becoming a researcher at the National Maritime Research Institute (in 2006), he is now an assistant professor in the Department of Logistics and Information Engineering, Tokyo University of Marine Science and Technology (since 2007). His research topics range from location analysis and logistics system modeling to transportation modeling.

Ikuho Yamada graduated from the University of Tokyo in 1997 and earned her PhD from University at Buffalo, the State University of New York, in 2004. After serving as an assistant professor at Indiana University–Purdue University, Indianapolis in 2004–2006, she is now an assistant professor in the Department of Geography, University of Utah. Her primary research interests reside in quantitative spatial analysis, GIScience, and their application to health- and transportation-related social issues.

Tohru Yoshikawa graduated from the University of Tokyo in 1985 and earned a doctor of engineering from the University of Tokyo in 1992. After becoming a research associate (in 1988) and assistant professor (in 1990) at Tokyo Metropolitan University, he is now an associate professor at Tokyo Metropolitan University (since 1993). His research topics range from urban spatial analysis to application of geographical information systems to city planning.

Section I

Urban Analysis and Planning Theories

Yasushi Asami

Urban analysis in Japan, originating from a lecture in the Department of Urban Engineering (founded in 1962) at the University of Tokyo, was initiated as a study of analytical tools in city planning. In the planning process, a number of quantitative evaluations have to be made, such as the prediction of population, land use demand, mobility pattern, and urban facility capacity. Urban analysis was regarded as an academic field serving such a need. To develop the tools further, to more sophisticated ones in the academic sense, researchers in urban analysis sought theoretical backgrounds from other (purer) scientific fields, such as economics, operations research, psychology, spatial statistics, etc., to be applied to urban phenomena.

The existing literature of urban analysis can be classified into several themes: (1) refinement of the measurement and evaluation of urban areas, (2) understanding the mechanism underlying actual urban phenomena, (3) better prediction of urban development, and (4) optimization of urban policies. The first three themes are rather positive analyses, where the major objective is to describe the actual situation in a better and useful manner. The five chapters in this part are dedicated to these themes. The last theme, on the other hand, is rather normative analysis, where the major objective is to propose the best urban policy based on some normative values.

The judgment in the planning process in many cases relies on the impression and intuition of experts, which is sometimes subject to mistake due to experts' lack of comprehension of the complicated interdependence and side effects of urban phenomena. Therefore, objective judgment based on urban analysis is important, even though the very detailed analyses are difficult. This applies to any of the themes stated above, and this is the *raison d'être* of urban analysis in the study of city planning.

Regarding the refinement of the measurement and evaluation of urban areas, one of the fundamental decisions should involve the appropriate selection of indices to measure certain aspects of urban space or urban activity. Planners often use indices to understand or express the characteristics of study areas properly. To do so, comparability among areas is one of the key features necessary for indices. Asami (in Chapter 1), taking an axiomatic approach, stresses the importance of checking the appropriateness of indices by considering the essential characteristics behind the functional form of the indices. To be more specific, the essence of ratio-type indices is summarized by the similarity axiom, which claims that if two areas of an equal index value are merged, then the integrated area should exhibit the same index value. One notable example is density measures such as population density, which is often used in planning. It is well known that population density may have different meanings if applied to areas of very different sizes. This indicates that even a simple measure of density has to be reexamined for appropriate usage in the planning process.

Regarding research for understanding the mechanism underlying actual urban phenomena, large volumes of literature have already been published. Typical topics include land use pattern, facility allocation, and human mobility. Studies in this theme start with a modest set of assumptions to mimic the actual urban phenomena. Theoretical derivation, numerical simulation, or fitting models to actual data leads to a certain set of results, which suggests some kind of tendency in urban phenomena. This linkage between a set of assumptions and their consequence is important information to the understanding of the urban phenomena, for the assumptions are regarded as causes and the consequence as their effect. If a certain consequence is robust in that the basic results do not change by modification of the set of assumptions (A), while it is subject to change by the modification of another set of assumptions (B), then we can judge that assumptions (B) are the fundamental ones leading to the consequence, while assumptions (A) are marginal and not so essential for the consequence. This kind of rule will become a precious guideline for experts in making planning decisions.

Tamagawa (in Chapter 2) started with the simple assumption that a new land use is allocated where it can best fit by maximizing affinity, which is a function of the local land use pattern. In particular, he examines the implication of micro-purification and macro-mixture. This is one of the established principles in land use planning. To protect the suitable environment for a certain land use, the land use should be pure by limiting other land uses to a small area. However, to ensure the benefits of interactions with other land uses, the land use pattern on a larger scale should be mixed. This principle is often applied to determine the allocation of land use zones. Tamagawa poses a question: What kind of spatial patterns will be generated through this principle? This seemingly easy question cannot be answered without Monte Carlo simulation, for actual interactions become very complicated. An amazing consequence is the long agglomeration pattern. In other words, narrow, linear (or curvy) development is the natural choice if agents of land uses allocate autonomously. This is rather counterintuitive. For example, do planners propose a long, narrow residential zone, instead of a thick clump? As is seen, urban analysis sometimes breaks the common sense of city planning, which is established by the naive intuition of experts.

Suzuki and Watanabe (in Chapter 3) analyzed the hierarchical allocation of facilities. Many urban facilities form a hierarchical system, such as the postal service, waste treatment facilities, administrative offices, education system, etc. What is the optimal hierarchical pattern for these facilities? Since this should heavily depend on the functional relationship among facilities as well as the consumption (or served) pattern of clients, the answer is not easy. With a simple objective to minimize the transportation cost among facilities, Suzuki and Watanabe derive analytically optimal hierarchical patterns of facilities that are subject to economies of scale. Their very important contribution is the derivation of a number of hierarchical levels, which are mostly assumed to be fixed in existing literature. Their notable device is the treatment of the integer problem as a continuous problem, which enabled the usage of the usual derivatives for optimization. Adopting a simple yet general enough assumption of the functional relationship related to economies of scale factors, they succeeded in deriving a general solution to the optimal level for hierarchy.

Kamachi, Asami, and Okabe (in Chapter 4) analyzed route-memorizing behavior. Nowadays, walking-through games such as dungeon games are popular on PCs or mobile PCs. When this chapter was developed, however, such screen simulation was one of the pioneering fields. Using such a device, route-memorizing experiments are conducted. Subjects can freely see maps when they walk through a maze-like street network. Under this circumstance, the walking route from the last consultation of the map to the next one can be regarded as the subjects' memory capacity for a route with a certain level of confidence. Based on this proposition, the memory load of route segments can be estimated. The results can be applied to estimate the best locations of guide maps on streets.

Finally, regarding research for better prediction of urban development, a vast amount of literature is devoted to the prediction of population, land use change, and transportation. To predict some numbers (such as future population), it is important to consider the stability of certain parameters over time. For example, when the population is predicted, stable parameters may be the survival rate and male/female ratio for newborn babies, while an unstable parameter may be the total fertility ratio. A more reliable prediction method should be based on stable parameters as much as possible. Sometimes it is very difficult to test the stability of parameters. In such cases, a cross-validation technique is often utilized. Typically acquired sample data are divided into two datasets, one for model building and the other for model testing. Prediction models are constructed based on the first dataset, and then the prediction accuracy of the model is tested with the other dataset. If the model is accurate enough for the other dataset, then the prediction method can be thought to be robust.

A number of models to predict land use change are proposed in the literature, but most of them are for predicting land use proportion in areas. Ito and Murata (in Chapter 5) developed a model that predicts point-wise land use change, since individual land use change occurs due to a variety of specific reasons, which are very difficult to predict. Accordingly, point-wise land use change is much harder to predict than land use proportion change in an area. Hence, increasing the accuracy of prediction for point-wise land use change is a big challenge. Ito and Murata succeeded in establishing a prediction model of a matching rate over 70% by using a flexible model of a neural network. Considering that they classified land use into six categories, this percentage can be said to be extraordinarily high.

Studies in city planning are apt for a detailed survey of specific cases, and as a result, the establishment of a general mechanism may become rather weak. Studies in urban analysis are conducted with the aspiration of a search for a general mechanism and general methods. Hence, urban analysis, if ideally studied, can complement the research in city planning in a favorable way. Five chapters in this section represent only a fragment of the achievement of urban analysis for planning. Nonetheless, these chapters potentially complement existing city planning literature with such an aspiration.

1 Characterization of Ratio-Type Indices for Evaluating Residential Environment*

Yasushi Asami[†]

1.1 INTRODUCTION

A variety of residential environment indices are calculated to survey the current situation of residential areas. Typical indices are ratio-type indices such as ratio and density. This is because absolute values such as population and area are not suitable for comparison owing to the difference in size of subjects' spatial units.

At least two features should be checked to justify the use of particular indices. First, the indices themselves must be appropriate. Inappropriate indices cannot convey sufficient information to judge the regional characteristics. Second, even if each index is appropriate, the index itself can only indicate a part of regional character. A comprehensive index that combines a set of indices should be realized by identification of an appropriate method. This second issue may depend on the purpose of the survey, and therefore the method of constructing a comprehensive index should take into account the particular survey aim. The first issue, however, can be a more general problem. One index can be used for a variety of purposes.

When we use ratio-type indices, how far should we check their appropriateness? We tend to assume that if the numerator and the denominator are well defined, then we understand the ratio-type index completely. But in this chapter, we question this presumption, and reconsider the kind of assumptions that are implicit in our use of ratio-type indices. Asami and Smith (1995) performed an axiomatic study of ratio-type indices, in which a set of axioms leading to a ratio-type index based on transaction matrix and final demand vector in input–output analysis was examined. The current chapter extends this study in two ways. First, it is shown that an additive-ratio index can be derived without assumption of the independence axiom that was necessary in Asami and Smith (1995), provided that the numerator and denominator are calculated from vectors. Second, it is shown that a simple sum-ratio-type index can be derived by the linearity-of-factor-2 and linearity-of-factor-3 axioms in addition to

* This is a translation of Asami (1996), with minor modifications to improve presentation.

† Department of Urban Engineering, the University of Tokyo (in 1995). Currently affiliated with the Center for Spatial Information Science, the University of Tokyo.

axioms for general sum-ratio-type indices. These two indices are not described in Asami and Smith (1995).

1.2 RATIO-TYPE INDEX FOR RESIDENTIAL ENVIRONMENT

To begin with, several indices are exemplified that are often used to evaluate a residential environment in the field of city planning:

1. Population density: This is given by population divided by the area of the target region. A typical population density is given by the population of residents divided by the area of the whole region. For example, an appropriate level of population density is said to be around 100 persons/ha for areas with detached housing, and areas with 200 persons/ha or more cannot preserve a proper residential environment unless some parts are converted into middle/high-rise apartments (Tsuchida et al., 1984). If population is replaced with daytime population, employment population for industries, population described by a specific attribute (such as a certain age group), or number of households, then the density becomes a measure for each specific aspect in a city.
2. Dwelling density: Instead of population, the number of dwelling units can be a numerator, which signifies physical density of houses, and this is often used in residential plans.
3. Population ratio: Measures such as the population ratio of young people, elderly people, and employers engaging in the primary industry are often used to convey the regional characteristics. For example, the population ratio of people more than or equal to 65 years old signifies the degree of aging of the particular society.
4. Household ratio: Instead of population ratio, household ratio may be used. For example, the ratio of households with elderly people and single-parent households are often used in housing policy.
5. Area ratio: In city planning, area ratio is also often used. For example, building coverage ratio and floor area ratio are fundamental regulation tools in the building standard law. Building coverage ratio is the ratio of building area to its site area, and floor area ratio is the ratio of total floor area to its site area. In Japan, their maximal values for appropriate residential areas are considered to be 60% for building coverage ratio and 200% for floor area ratio.

In addition to these ratios, building use ratios such as residential use ratio (total floor area for residential use divided by total floor area), industrial use ratio (total floor area for industrial use divided by total floor area), wooden building ratio (total floor area of wooden buildings divided by total floor area), and wooden apartment ratio (total floor area of wooden apartment houses divided by total floor area) are often used to characterize the areas (Tsuchida et al., 1984).

As regards the ratio-type measures described above, there are basically two types. One is the density measure, where the unit of the numerator and that of the denominator are different, and the other is the ratio measure, where the unit of the numerator and that of the denominator are the same, such as the ratio of floor areas. The density measures require special attention, as described in this chapter's conclusion.

1.3 AXIOMATIC APPROACH

For surveying the residential environment of a region, the indices to be surveyed are chosen to fit the aspects to be focused on. Suppose that we are interested in how dense buildings are located. We would expect that the larger the buildings in a region with fixed area, the higher the index should be. We also expect that if the volumes of buildings are the same, the larger the area of the region, the smaller the index should be. If the index is defined only by the total floor area of buildings and area of the region, then floor area ratio seems to be the right choice for the index. There are more indices, however, that satisfy the characteristics mentioned above. An index defined as total floor area minus area of region and another index defined as square of total floor area divided by area of region would also satisfy the requirement above. This shows that the characteristics permit much more variety of indices. Then why do we prefer floor area ratio index to other indices? This is because we implicitly assume other conditions desirable for an index that are not satisfied by other indices.

What would be the complete set of conditions for an index leading to a ratio-type measure? The axiomatic approach can solve this question. That is, a set of axioms stating the desirable conditions for an index is tested to see if they can confine the functional form of the index to ratio-type.

1.4 CHARACTERISTICS OF RATIO-TYPE INDICES

Consider the floor area ratio index. Let p be total floor area and s (>0) be the area of the region. Assume that the index to be derived is expressed as a function, $f(p, s)$, of p and s. The problem here is the determination of characteristics of function, f, such that $f(p, s)$ is expressed as

$$f(p, s) = c(p/s) \tag{1.1}$$

where c is a positive constant, or its extensive form

$$f(p, s) = \phi(p/s) \tag{1.2}$$

where $\phi(.)$ is monotone increasing function.

The following is a list of characteristics that floor area ratio satisfies:

1. An infinitesimal change in total floor area or area of the region brings about only an infinitesimal change in floor area ratio.
2. If total floor area increases, floor area ratio increases.
3. If two regions with the same floor area ratio are combined into one region, the floor area ratio does not change.

4. If total floor area is zero, then the floor area ratio takes the minimum value irrespective of the area of the region.
5. If the area of the region increases while holding the total floor area constant, then the floor area ratio decreases.
6. If two regions with different floor area ratio are combined into one region, then the floor area ratio becomes a value between the two original values.
7. If the total floor area is doubled, then the floor area ratio is doubled.
8. If the area of the region is halved while holding the total floor area constant, then the floor area ratio is doubled.
9. If the total floor area is tripled, then the floor area ratio is tripled.

The characteristics mentioned above are only part of the floor area ratio index. This can easily be checked by the fact that any single item from (1) to (9) is not enough to determine the functional form of floor area ratio in Equation (1.1) or (1.2).

1.5 SET OF AXIOMS AND INDICES

Do all the items from (1) to (9) determine the functional form? It appears that the indices satisfying all the characteristics should be of the same form as Equation (1.1). Moreover, the indices satisfying characteristics from (1) to (6) should be of the same form as Equation (1.2).

The next question is if all the characteristics from (1) to (6) are necessary to confine the indices of the form (1.2). The answer is no. We only need characteristics from (1) to (3). In other words, these three characteristics are the essential feature of ratio-type indices. Please refer to Appendix 1.1 for the proof. Here the implication of this result will be discussed. Let R, R_+, R_{++} be a set of real numbers, a set of nonnegative real numbers, and a set of positive real numbers, respectively. The index function is defined by f: $R_+ \times R_{++} \rightarrow R$.

Continuity axiom: If p or s changes infinitesimally, the index changes infinitesimally. More rigorously, for any series (p_m, s_m) converging to (p, s), $f(p_m, s_m)$ converges to $f(p, s)$.

Monotonicity axiom: If p increases, then the index increases. More rigorously, $p_1 < p_2$ implies $f(p_1, s) < f(p_2, s)$.

Similarity axiom: If two regions with the same index value are combined, the index value does not change. More rigorously, $f(p_1, s_1) = f(p_2, s_2)$ implies $f(p_1 + p_2, s_1 + s_2) = f(p_1, s_1)$.

Characterization theorem for general ratio-type indices: If f: $R_+ \times R_{++} \rightarrow R$ satisfies continuity, monotonicity, and similarity axioms, then f can be expressed as $f(p, s) = \phi(p/s)$, where ϕ is a monotone increasing function.

Continuity axiom means that the index changes continuously with p and s without drastic change, such as jump. Continuity is a feature that is often implicitly assumed for many indices, and hence this requirement may be said to be a natural one.

Monotonicity axiom means that the larger the total floor area, the larger the index, provided that the area of the region does not change. This is *the* feature required for indices to signify the level of congestion of buildings in a region.

1.6 SIGNIFICANCE OF SIMILARITY AXIOM

Similarity axiom means that the index value does not change if two regions of the same index value are combined. This similarity axiom is the essential axiom leading to ratio-type indices, and therefore its implication should be clearly understood.

Consider the case where we calculate floor area ratio for different sizes of urban districts. If the district consists of an urban block, then it does not contain sites other than building sites. If the district consists of a neighborhood unit (its size being equivalent to an elementary school zone), then it may contain a neighborhood park, parks for children, and space for roads. This means that the denominator includes sites other than building sites. If the district is large enough to include several neighborhood units, then regional parks, space for railways, sites for facilities for water, sewage treatment, waste disposal and treatment, etc., will also be included. If the district extends to the city itself, then a city park, river, harbor, agricultural land, and forests will be included.

This means that the floor area ratio tends to be smaller when the area of the district becomes larger. If regions of the same floor area ratio are combined, then in reality the floor area ratio will decrease. The reason why this logically odd phenomenon is observed is that the (spatially connected) region tends to include nonbuilding sites in reality. Of course, if all these sites are ignored, then the similarity axiom holds in reality as well.

Of the three axioms, continuity and monotonicity seem natural requirements. The similarity axiom can therefore be regarded as the essential axiom characterizing the ratio-type indices. It follows that when we are going to use ratio-type indices, we need to check their validity as regards the similarity axiom, and if it is satisfactory, then we can safely use the index.

1.7 SIMPLE RATIO-TYPE INDICES

It was shown above that only three axioms are needed to characterize the generalized ratio-type form as in Equation (1.2). To confine the indices to the form (1.1), additional conditions are necessary.

For example, addition of condition (7) may appear sufficient to get functional form (1.1), but it appears that this condition is not enough. Interestingly enough, the further addition of condition (9), which looks very similar to (7), now works well to confine the indices of the form (1.1). The conditions (7) and (9) can be expressed as the following axioms:

Linearity-in-factor-2 axiom: If p is doubled, then the index is doubled.
That is, $f(2p, s) = 2f(p, s)$ holds.
Linearity-in-factor-3 axiom: If p is tripled, then the index is tripled.
That is, $f(3p, s) = 3f(p, s)$ holds.

Use of both axioms implies that if total floor area becomes a certain number times, then the index becomes this value times. Thus, this property means that the index is ratio-scale measure (Tanaka, 1977). This also implies that the ratio of index value carries important meaning, and in particular the meaning of zero is very critical. For the case of floor area ratio, zero means that there is no building at all, and this state is in a sense a starting point for measuring floor area ratio. Both linearity-in-factor-2 axiom and linearity-in-factor-3 axiom regulate the relationship for ratio of specific factor. If continuity, linearity-in-factor-2, and linearity-in-factor-3 axioms are put together, then the index has to satisfy the following linearity in real numbers axiom, as shown in Appendix 1.2:

Linearity in real numbers axiom: For any positive real number, r, if p becomes r times, then the index becomes r times. That is, for any real number, r, $f(rp, s) = rf(p, s)$ holds.

Characterization theorem for simple ratio-type indices: If f: $R_+ \times R_{++} \rightarrow R$ satisfies continuity, monotonicity, similarity, linearity-in-factor-2, and linearity-in-factor-3 axioms, then f can be expressed as $f(p, s) = c(p/s)$, where c is a positive constant.

1.8 SUM-RATIO-TYPE INDICES

When numerator and denominator are expressed as linear combinations, the index will be called a sum-ratio-type index. When we calculate the floor area ratio for a region, the actual procedure is to measure the floor area and area of subregions followed by a summing up of these values to get total floor area and the area of the region. Thus, the floor area ratio itself is a sum-ratio-type index.

To proceed with the discussion: the following notations are introduced. Regions are indicated by $i = 1, \ldots, n$. Total floor area for region i is denoted by p_i and its area by s_i (>0). Since the index is now dependent on the vector of floor area, $p = (p_i)$, and vector of area of regions, $s = (s_i)$ $(i = 1, \ldots, n)$, the index is expressed as $f(p, s)$. A simple sum-ratio-type index is defined as

$$f(p, s) = c(p_1 + p_2 + \ldots + p_n)/(s_1 + s_2 + \ldots + s_n) \quad (1.3)$$

where c is a positive constant, and a general sum-ratio-type index is defined as

$$f(p, s) = \phi[(p_1 + p_2 + \ldots + p_n)/(s_1 + s_2 + \ldots + s_n)] \quad (1.4)$$

where $\phi[.]$ is a monotone increasing function.

For sum-ratio-type indices, we can determine a set of conditions leading to the functional form of Equations (1.3) and (1.4). To do so, continuity, monotonicity, and similarity axioms are described to fit in the case of n regions. In doing this, we use the following vector expression:

$$x > y \text{ if and only if for any } i \ (=1, \ldots, n), x_i > y_i$$

$$|x| = (x_1^2 + \ldots + x_n^2)^{1/2}$$

Superscript t stands for transposition operator.

$$\mathbf{1} = (1, \ldots, 1)^t$$

The index is defined by f: $R^n_+ \times R^n_{++} \to R$.

Continuity axiom: If p or s changes infinitesimally, then the index changes infinitesimally as well. More rigorously, for any series (p_m, s_m) converging to (p, s), $f(p_m, s_m)$ converges to $f(p, s)$.

Monotonicity axiom: If p increases, then the index increases. More rigorously, $p_1 < p_2$ implies $f(p_1, s) < f(p_2, s)$.

Similarity axiom: If two regions with the same index value are combined, the index value does not change. More rigorously, $f(p_1, s_1) = f(p_2, s_2)$ implies $f(p_1 + p_2, s_1 + s_2) = f(p_1, s_1)$.

Just these three axioms together cannot confine the index to be of the form (1.4). For example, supposing that a_i, b_i are positive real number constants, $f(p, s) = (\Sigma_i a_i p_i)/(\Sigma_i b_i s_i)$ also satisfies these axioms. For the sake of equal weights for all the regions, the condition is necessary that the order of regions does not influence the value of index. Let E_{ij} be a matrix given by exchanging the i-th and j-th rows of identity matrix, I. To exchange region i and region j, we only need to multiply E_{ij} to p and s from the left-hand side. The set of exchanging matrix, E_{ij}, is denoted by E. The condition that exchanging regions does not influence the value of index can be summarized in the symmetry axiom.

Symmetry axiom: The order of region does not influence the value of the index. That is, for any E_{ij} in E, $f(E_{ij}p, E_{ij}s) = f(p, s)$ holds.

Symmetry axiom means that changing the order of regions does not affect the value of index, which is also implicitly assumed for many indices. Continuity, monotonicity, similarity, and symmetry axioms imply that the index is of form (1.4) (for the proof, see Asami (1996)).

Characterization theorem for general sum-ratio-type indices: If f: $R^n_+ \times R^n_{++} \to R$ satisfies continuity, monotonicity, similarity, and symmetry axioms, then f can be expressed as $f(p, s) = \phi[(p_1 + p_2 + \dots + p_n)/(s_1 + s_2 + \dots + s_n)]$, where ϕ is a monotone increasing function.

This characterization theorem for general sum-ratio-type indices together with the discussion above implies that even for sum-ratio-type indices the essential driving force leading to the ratio-type is the similarity axiom. Accordingly, the argument in Section 1.6 applies here, too.

1.9 SIMPLE SUM-RATIO-TYPE INDICES

Four axioms are shown to be sufficient to characterize general sum-ratio-type indices. To confine the indices to a simple sum-ratio-type, additional conditions are necessary again.

Additional axioms introduced for the simple ratio-type indices are modified for the case of n region in the following:

Linearity-in-factor-2 axiom: If p is doubled, then the index is doubled. That is, $f(2p, s) = 2f(p, s)$ holds.

Linearity-in-factor-3 axiom: If p is tripled, then the index is tripled. That is, $f(3p, s) = 3f(p, s)$ holds.

The addition of these two axioms is enough to characterize the simple sum-ratio-type indices. (Refer to Asami (1996) for the proof.)

Characterization theorem for simple sum-ratio-type indices: If f: $R^n_+ \times R^n_{++} \rightarrow R$ satisfies continuity, monotonicity, similarity, symmetry, linearity-in-factor-2, and linearity-in-factor-3 axioms, then f can be expressed as $f(p, s) = c[(p_1 + p_2 + \ldots + p_n)/(s_1 + s_2 + \ldots + s_n)]$, where c is a positive constant.

1.10 CONCLUSION: RATIO-TYPE AND SUM-RATIO-TYPE INDICES RECONSIDERED

The argument for the characterization theorems above is based on the implicit assumption that both numerator and denominator are real-valued. For some ratio-type indices for residential environments, this does not hold. For example, the numerator is always integer-valued for population density, and the numerator and denominator are integer-valued for population ratios for certain attributes. In such cases, the continuity axiom may not be meaningful. Even if the numerator or denominator of ratio-type indices is integer-valued, we tend to expect that the index satisfies the continuity axiom by behaving as if real values are possible. The results above can be extended to such cases.

In Section 1.6, it was stated that floor area ratio tends to become smaller if the area of the region is enlarged. In the practice of city planning, the size of the region is always important information, even when density type indices are used (and hence standardization is implicitly assumed). A similar argument holds for population density. If density type indices are used, and if the size of the region is necessary, then this implies that the standardization of density type indices is not perfect.

City planners are devising density type indices to control the standardization issue. The density indices described in Section 1.2 are called gross density. For gross density, the denominator counts up areas irrespective of the fact that the subregion may not accommodate the target activity counted for the numerator. In contrast, there are density indices called net density. For net density, the denominator counts up only subregions that can accommodate the target activity counted for the numerator.

For example, if area of habitable region is used in calculating population density, then the index becomes more comparable for regions of different size. In particular, if the area for public facilities and agriculture, which cannot be used for housing sites, is deducted, then the density becomes net density. Net density becomes higher value than gross density, and it signifies actual degree of population congestion in residential areas. When floor area ratio is calculated, if the denominator includes only sites for buildings, then it becomes net floor area ratio. Net coverage ratio can be devised similarly.

The net densities are not complete, however. The possibility of placing the target activities varies from site to site, and hence the net density can become arbitrary, depending on the definition of counting the valid area.

Careful examination is necessary in selecting density type indices to reflect the meaning of the index. This chapter clarifies that checking the validity of the similarity axiom is the key in adopting ratio-type (or sum-ratio-type) indices. If the similarity axiom is judged appropriate, then the use of ratio-type (or sum-ratio-type) indices is justified and the meaning of the indices is favorably understood.

REFERENCES

Asami, Y. 1996. *Characteristics of the ratio-type indices for residential environment.* Discussion Paper 65, Department of Urban Engineering, University of Tokyo.

Asami, Y., and Smith, TE. 1995. Additive-ratio measures of interactivity in input-output systems. *Journal of Regional Science* 35:85–115.

Tanaka, Y. 1977. *Psychological measurement methods.* Tokyo: University of Tokyo Press.

Tsuchida, H., Itami, M., Hibata, Y., Uchida, Y., Hayashi, Y., and Takamizawa, K. 1984. *Urban development and improvement planning. New series in architectural studies 19.* Tokyo: Shokokusha.

APPENDIX 1.1: OVERVIEW OF PROOF OF CHARACTERIZATION THEOREM FOR GENERAL RATIO-TYPE INDICES (FOR RIGOROUS PROOF, SEE ASAMI (1996))

Similarity axiom together with mathematical induction implies that for any positive integer, k, $f(kp, ks)$ holds. Since we have $f(p, s) = f(kp, ks) = f(m(k/m)p, m(k/m)s) = f((k/m)p, (k/m)s)$ for a positive real number m, for any positive rational number, q, $f(qp, qs) = f(p, s)$ holds. It follows from the denseness of rational numbers in real number space and continuity axiom that for any positive real number, r, $f(rp, rs) = f(p, s)$ holds. By taking $r = 1/s$, $f(p, s) = f(p/s, 1)$. By letting $\phi(x) = f(x, 1)$, $f(p, s) = \phi(p/s)$. Monotonicity axiom implies that $f(.)$ is a monotone increasing function.

APPENDIX 1.2: OVERVIEW OF PROOF OF CHARACTERIZATION THEOREM FOR SIMPLE RATIO-TYPE INDICES (FOR RIGOROUS PROOF, SEE ASAMI (1996))

By characterization theorem for generalized ratio-type indices, continuity, monotonicity, and similarity axioms imply that the functional form should be as expressed in Equation (1.2). It follows from linearity-in-factor-2 and linearity-in-factor-3 axioms that for any nonnegative real number, x, and integer, a, b, $\phi(2^a 3^b x) = 2^a 3^b \phi(x)$ holds. Since $\{Y = 2^a 3^b$: a, b are integers$\}$ is dense in nonnegative real number space, continuity axiom implies that for any nonnegative real number, r, $\phi(rx) = r\phi(x)$ holds. By letting $c = \phi(1)$, Equation (1.1) holds.

2 A Compound Simulation Model of Land Use Patterns and Its Implications*

Hidenori Tamagawa

2.1 INTRODUCTION

Morphological assessment of the land use patterns of built-up areas is an element in the fields of urban and regional analysis that has been researched for many years. Mesh data of land use patterns have been studied for over 30 years.

Studies by Koide (1977), Tamagawa (1982), Sugita and Koshizuka (1983), and Yoshikawa (1997, 1998, 1999), for example, have focused on the degree of mixed use, agglomeration, and adjacency of uses to examine the characteristics of urban patterns. A study by Gatrell (1977) is an attempt to relate the amount of information (the degree of redundancy in particular) and the characteristics of patterns, including agglomeration and adjacency, to spatial lag. These are so-called descriptive models that emphasize reproducibility. Their effectiveness has been evaluated by how closely they conform to patterns in the real world.

On the other hand, there are alternative methods of analysis, i.e., those that deduce urban patterns based on a prescriptive model with predetermined evaluation standards and extract useful information while monitoring the results of the simulation. For example, Portugali and colleagues (1994) modeled the social phenomenon of segregation using cellular dynamics and examined both stability and instability within a city. This method of study, however, has not been conducted in Japan since the classical work by Okudaira (1976), who conducted an experiment on land use patterns. His study presented several interesting patterns generated by simulations, but did not systematically examine them.

This chapter is an attempt to take a step forward from Okudaira's (1976) study by examining a compound simulation to create a model of the "orderly complexity" of land use. As explained below, *compound* refers to the combination of relationships between uses at two levels and the implications of simulation results for a planning

* This is a translation of Tamagawa (2000), with minor modifications to improve presentation.

paradigm (purification of uses at the micro-level and a greater mixture of uses at the macro-level).

2.2 METHOD OF RESEARCH

The outline of Monte Carlo simulation used in the study to simulate the land use pattern is as follows.

2.2.1 Premise

Create a grid pattern consisting of 100 cells with 4 different land use categories (each colored differently and generated in the ratio of 4:3:2:1 to allow a mixed use of different probabilities) and generate 100 grid patterns per parameter.

2.2.2 Procedure

1. Determine the use and location of the first cell ($i = 1$) by random selection.
2. The use of the ith cell ($i \geqq 2$) is selected randomly and its location is based on the optimal criteria. Optimal criteria are those that maximize the total affinity generated by interaction between different uses between the ith cell and the cells already generated ($j < i$):

$$\sum_{\substack{j<i \\ d_{ij} \leq r_1}} \frac{a_{LU(i)LU(j)}}{d_{ij}^k} + \sum_{\substack{j<i \\ r_1 < d_{ij} \leq r_2}} \frac{b_{LU(i)LU(j)}}{d_{ij}^k} \rightarrow \max \tag{2.1}$$

under the following conditions:

$LU(i)$: Land use category of the ith cell
d_{ij}: Distance between the ith cell and the jth cell
k: A coefficient expressing the influence of distance, i.e., the resistance to distance
r_1, r_2: Extent of the impact of affinity
a_{**}, b_{**}: Coefficients indicating interaction between land use categories

This is a type of gravity model using three additional parameters: (1) resistance to distance, (2) extent of the impact, and (3) interaction (details explained in (4) and (5) below). To enhance Okudaira's (1976) simulation further, two levels indicating the extent of the impact of affinity were set. Location of uses is determined by the affinities. In other words, the model is a compound simulation model taking into account the relationship between cells.

This process is repeated ($2 \leqq i \leqq 100$) to generate a single land use pattern of a grid of 100 cells.

3. Next, the initial setting of the random number is changed using the same parameter setting to generate 100 different patterns, and the statistical values are calculated as explained below.

4. Parameters that affect the distribution, i.e., resistance to distance k, and the extent of the impact of affinity r_1, r_2 are set from the following options (provided that distance between the adjacent cells is set equal to 1):

$$k = 0, 2$$

$$r_1 = 1 \qquad r_1 = 1 \qquad r_1 = \sqrt{2}$$

$$r_2 = 2 \qquad r_2 = 2\sqrt{2} \qquad r_2 = 2\sqrt{2}$$

Here, two types of k—(i) where resistance to distance is zero and (ii) where resistance to distance is as it is in the ordinary gravity model—and three sets of r_1 and r_2 (to limit the extent of the impact to vertical or horizontal adjacent cells or to include diagonal cells) are chosen. To avoid the boundary value problem (judgment errors), slightly larger numbers (1.1 for 1, 2.1 for 2, 1.5 for sqrt(2), and 2.9 for 2* sqrt 2) have been input.

5. Coefficients indicating interaction between uses (a_{**}, b_{**} are aligned as matrices A and B) in this study are as follows: (i) no interaction exists between uses in matrix O, (ii) interaction only exists between same uses in matrix I, and (iii) interaction only exists between different uses in matrix N:

$$O = \begin{matrix} 0 & 0 & 0 & 0 \\ 0 & 0 & 0 & 0 \\ 0 & 0 & 0 & 0 \\ 0 & 0 & 0 & 0 \end{matrix}, \quad I = \begin{matrix} 1 & 0 & 0 & 0 \\ 0 & 1 & 0 & 0 \\ 0 & 0 & 1 & 0 \\ 0 & 0 & 0 & 1 \end{matrix}, \quad N = \begin{matrix} 0 & 1 & 1 & 1 \\ 1 & 0 & 1 & 1 \\ 1 & 1 & 0 & 1 \\ 1 & 1 & 1 & 0 \end{matrix}$$

They are used in the following five types of combination:

a. A = B = O: This corresponds to a so-called random pattern and will be indicated by the term *Rnd* in Table 2.1.
b. A = I and B = O: Indicated by I_O, assuming that interaction exists only between same uses within the extent of the impact r_1. This is a pattern of purification of uses at the micro-level (micro-purification).
c. A = O and B = N: Indicated by O_N, assuming that interaction exists between different uses and only within r_1 and r_2. In other words, this is more of a mixture of uses at a macro-level (macro-mixture).
d. A = I and B = N: Indicated by I_N. This is a combination of (b) and (c) above or a combined pattern of micro-purification and macro-mixture.
e. A = I and B = N but b_{**} operates not within $r_1 < d \leqq r_2$ but between $d \leqq r_2$ in formula (2.1). Indicated by I_N_{ov}, the potential for mixture is overlaid at the micro-level in type (d) above.

In this simulation, rather than generating random patterns in all cells and then optimizing them, the location of newly generated uses in each process is optimized as shown in procedure 2 of Section 2.2.2 for the following two reasons: (1) it more

TABLE 2.1
Indices of Random Patterns

r	LU	Data	
Rnd	1	Average: J	28.57
		Average: Cy	1.98
		Average: Cl	13.41
	2	Average: J	15.61
		Average: Cy	0.6
		Average: Cl	14.99
	3	Average: J	6.91
		Average: Cy	0.05
		Average: Cl	13.14
	4	Average: J	1.66
		Average: Cy	0.01
		Average: Cl	8.35

closely resembles the actual process of city formation, and (2) it takes less time to execute the program. Because the genetic algorithm (GA) approach, which is often used in solving optimization problems, was difficult to apply in this case, the optimal standard itself was simplified to make it easier to determine the location of the generated cells. It will be an interesting topic for future study to investigate how these differences affect the formation of patterns.

2.3 RESULTS OF THE SIMULATION

To examine the morphological characteristics of patterns obtained by the Monte Carlo simulation, two new figures are used: "cycle" (Cy in Tables 2.1 through 2.3) of the same uses and "clump" (Cl in the tables) of the same uses. These indicate a higher level of agglomeration than "join" (J in the tables) used in the descriptive model of Koide (1977), for example (Nota bene: Because Euler's formula, i.e., number of joins minus number of cycles plus number of clumps = number of cells, is true in this case, there is an overlap of information).

The degree of agglomeration is judged by the degree of deviation from randomness. A significant deviation from randomness is determined by a statistical test between one hundred samples obtained from simulation (testing the difference between population means by the corresponding sample) rather than theoretically calculated expected values or rank values (or class values) using diffusion. This is to remove the possible bias owing to randomizing.

Table 2.1 shows the average numbers of join, cycle, and clump in one hundred samples of random patterns. Table 2.2 shows the figures of I_O, O_N, and I_N, and Table 2.3 of $I\_N_{ov.}$ The test in Table 2.2 was conducted between I_N and the random pattern, and that in Table 2.3 (calculating *t* value) was conducted between I_N_{ov} and the random pattern.

TABLE 2.2
Various Indices of Patterns with Interactions

			Upper Column: Resistance to Distance (*k*)				Lower Column: Interaction Matrix A_B			
			k = 0				*k* = 2			
r1_r2	LU	Data	A_B = I_0	0_N	I_N	*t* Value (against Random)	I_0	0_N	I_N	*t* Value (against Random)
1.1_2.1	1	Average: J	60.92	28.24	36.49	19.08	60.37	28.05	48.04	38.19
		Average: Cy	22.45	1.27	4.38	10.28	21.94	0.65	10.91	24.18
		Average: Cl	1.53	13.03	7.89	−17.57	1.57	12.60	2.87	−36.92
	2	Average: J	43.31	15.87	23.13	21.88	42.75	15.77	33.06	44.49
		Average: Cy	14.91	0.20	2.09	10.84	14.38	0.06	6.65	23.60
		Average: Cl	1.60	14.33	8.96	−20.19	1.63	14.29	3.59	−43.18
	3	Average: J	27.25	6.97	11.87	20.71	26.83	6.95	19.20	40.17
		Average: Cy	8.86	0.04	0.63	8.48	8.42	0.01	3.04	15.32
		Average: Cl	1.61	13.07	8.76	−20.29	1.59	13.06	3.84	−42.37
	4	Average: J	11.27	1.66	4.08	14.43	10.79	1.72	7.33	26.47
		Average: Cy	2.87	0.00	0.06	1.91[a]	2.53	0.00	0.78	8.92
		Average: Cl	1.60	8.34	5.98	−14.37	1.74	8.28	3.45	−26.80
1.1_2.9	1	Average: J	60.92	37.10	40.96	29.22	60.37	33.29	44.45	35.67
		Average: Cy	22.45	3.94	6.65	18.09	21.94	1.61	8.12	19.20
		Average: Cl	1.53	6.84	5.69	−25.60	1.57	8.32	3.67	−32.35
	2	Average: J	43.31	22.72	26.13	31.93	42.75	20.17	30.56	44.00
		Average: Cy	14.91	1.39	2.98	16.60	14.38	0.47	4.84	21.39
		Average: Cl	1.60	8.67	6.85	−29.15	1.63	10.30	4.28	−41.71
	3	Average: J	27.25	10.80	13.55	27.84	26.83	9.49	17.65	41.06
		Average: Cy	8.86	0.19	0.97	11.15	8.42	0.10	2.17	15.05
		Average: Cl	1.61	9.39	7.42	−27.25	1.59	10.61	4.52	−42.18
	4	Average: J	11.27	2.72	4.40	14.55	10.79	2.25	6.81	26.26
		Average: Cy	2.87	0.00	0.17	3.81	2.53	0.01	0.60	7.83
		Average: Cl	1.60	7.28	5.77	−14.62	1.74	7.76	3.79	−27.48
1.5_2.9	1	Average: J	61.21	38.81	45.57	37.02	61.65	38.04	58.34	57.85
		Average: Cy	22.85	5.86	10.72	35.52	23.16	5.66	20.12	51.30
		Average: Cl	1.64	7.05	5.15	−27.02	1.51	7.62	1.78	−40.23
	2	Average: J	43.73	23.67	30.50	40.03	43.80	23.36	40.85	54.44
		Average: Cy	15.48	2.57	6.24	31.03	15.42	2.65	13.02	36.82
		Average: Cl	1.75	8.90	5.74	−34.72	1.62	9.29	2.17	−47.25
	3	Average: J	27.21	11.33	17.17	36.81	27.69	11.20	24.67	54.12
		Average: Cy	8.87	0.56	3.02	29.71	9.23	0.78	6.90	33.63
		Average: Cl	1.66	9.23	5.85	−32.48	1.54	9.58	2.23	−53.34
	4	Average: J	11.14	2.75	5.76	17.80	11.59	2.75	9.22	32.23
		Average: Cy	2.81	0.06	0.57	7.83	3.03	0.02	1.64	13.74
		Average: Cl	1.67	7.31	4.81	−18.54	1.44	7.27	2.42	−38.86

[a] This figure only is 5% significant and not 1%. Other figures are all 0.1% significant.

TABLE 2.3
Various Indices of Patterns with Overlay of Interaction

			Upper Column: Resistance to Distance (*k*)			Lower Column: Interaction Matrix A_B		
			k = 0			*k* = 2		
r1_r2	LU	Data	I_Nov	*t* Value (against Random)	Judgment	I_Nov	*t* Value (against Random)	Judgment
1.1_2.1	1	Average: J	29.84	3.10	***	29.20	1.53	X
		Average: Cy	1.51	−2.39	**	1.10	−4.66	***
		Average: Cl	11.67	−5.58	***	11.90	−4.81	***
	2	Average: J	16.92	3.86	***	16.64	2.87	**
		Average: Cy	0.24	−3.60	***	0.20	−4.40	***
		Average: Cl	13.32	−5.65	***	13.56	−4.42	***
	3	Average: J	7.86	4.53	***	7.30	1.64	X
		Average: Cy	0.02	−1.35	X	0.03	−0.71	X
		Average: Cl	12.16	−4.70	***	12.73	−1.75	*
	4	Average: J	2.06	2.24	*	1.82	0.94	X
		Average: Cy	0.00	XXX		0.00	XXX	
		Average: Cl	7.94	−2.32	*	8.18	−1.01	X
1.1_2.9	1	Average: J	37.26	21.55	***	31.95	7.96	***
		Average: Cy	4.11	9.89	***	1.86	−0.62	X
		Average: Cl	6.85	−22.05	***	9.91	−11.06	***
	2	Average: J	22.23	18.50	***	18.59	8.81	***
		Average: Cy	1.45	6.19	***	0.43	−1.61	X
		Average: Cl	9.22	−18.91	***	11.84	−10.84	***
	3	Average: J	10.50	13.86	***	8.54	6.61	***
		Average: Cy	0.30	4.22	***	0.05	0.00	X
		Average: Cl	9.80	−14.11	***	11.51	−6.80	***
	4	Average: J	2.41	4.15	***	2.25	3.18	***
		Average: Cy	2.00	XXX		0.00	XXX	
		Average: Cl	7.59	−4.27	***	7.75	−3.28	***
1.5_2.9	1	Average: J	36.80	16.82	***	32.02	6.93	***
		Average: Cy	5.17	13.54	***	3.27	5.82	***
		Average: Cl	8.37	−14.51	***	11.25	−5.84	***
	2	Average: J	22.69	17.52	***	18.73	7.87	***
		Average: Cy	2.21	9.68	***	1.19	4.35	***
		Average: Cl	9.52	−17.45	***	12.46	−7.71	***
	3	Average: J	10.41	11.44	***	8.21	4.60	***
		Average: Cy	0.63	8.30	***	0.22	3.31	***
		Average: Cl	10.22	−10.56	***	12.01	−4.30	***
	4	Average: J	2.25	2.82	**	1.95	1.58	X
		Average: Cy	0.04	1.35	X	0.02	0.58	X
		Average: Cl	7.79	−2.78	**	8.07	−1.57	X

XXX, judgment impossible with the sample variance = 0; X, not 5% significant; *, 5% significant but not 1%; **, 1% significant but not 0.1%; ***, 0.1% significant.

We can see from the figures in Table 2.2 that a compound pattern of I_N is situated between I_O and the random pattern, and between I_O and O_N. Compared with the random pattern, I_O has higher numbers of join and cycle, indicating a clear tendency to agglomeration. The numbers of cycle, however, did not increase as much as the numbers of join. This tendency is clearly seen among uses of small proportion. As most results in the table are significant at the 0.1% level, the *t* values are listed as an index indicating the degree of agglomeration compared with random patterns. These figures show a similar tendency, as in the case of figures for join and cycle.

Effects of each parameter are as follows:

1. Extent of the impact: The wider the agglomeration, the stronger the tendency to agglomeration. In the case of $(r_1, r_2) = (1.5, 2.9)$ in particular, the degree of agglomeration of join and cycle is closer.
2. Effect of the resistance to distance: The greater the pattern's resistance to distance is, the greater the degree of agglomeration. This is because the greater the distance, the lower the effect of the mixture.
3. Effect of overlay (Table 2.3): Compared with I_N, I_N_{ov} has a greater divergence between join and cycle. The extent of the impact was similar to that without overlay. In the case of $(r_1, r_2) = (1.1, 2.1)$ with narrow impact (compared with the case of 1.5, 2.9 above) in particular, while join emerged on the agglomeration side, the numbers of cycle in some cases were statistically significantly smaller than those of random pattern. The effect of the resistance to distance was slightly negative in the case of "with overlay" compared with "without overlay."

Figures 2.1 to 2.12 show the examples of patterns obtained. Figures 2.1 to 2.6 are the result of one random sequence, and Figures 2.7 to 2.12 are the result of another random sequence. The figures show how the patterns change depending on the interaction between uses. These figures have been chosen from 100 samples whose sum of squared deviations of the numbers of join, cycle, and clump (the sum of four uses) was of the smallest ten. These therefore show typical patterns with all three indices close to the average under their conditions. The top row of each column shows micro-purification patterns, the middle row shows micro-purification and macro-mixture mixed patterns, and the bottom row shows micro-purification and macro-mixture patterns, which allows the overlay. Long agglomerative patterns are observed in the middle row if focused on uses with a higher ratio (black circles and white circles). This tendency is slightly stronger with a pattern of larger resistance to distance ($k = 2$). On the other hand, in patterns allowing the overlay the uses are more fragmented, rather than showing the tendency of agglomeration.

2.4 IMPLICATIONS

A very general observation of the simulation outcome is that under a certain condition the logical processes of micro-purification and macro-mixture tend to generate long agglomerative patterns. The "certain condition" in the simulation in this

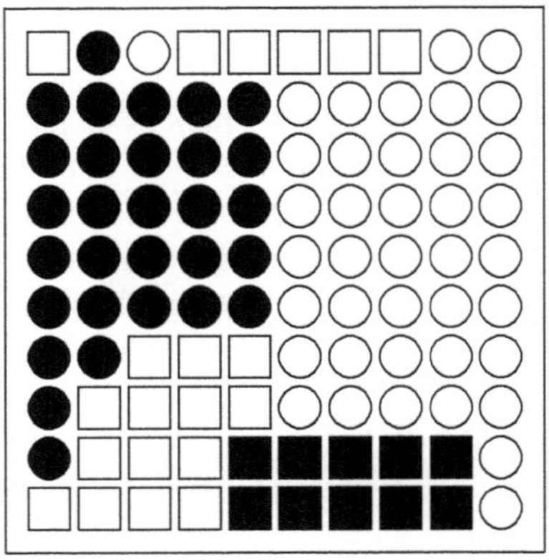

FIGURE 2.1 Typical example of patterns as a result of simulation. $r_1 = 1.1$, $k = 0$, I_0.

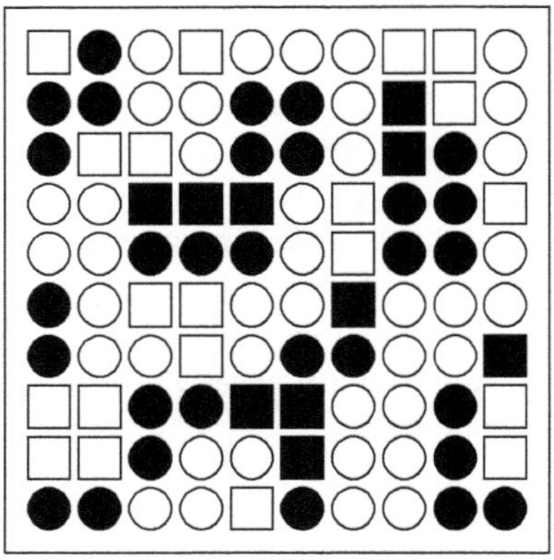

FIGURE 2.2 Typical example of patterns as a result of simulation. $r_1 = 1.1$, $r_2 = 2.1$, $k = 0$, I_N.

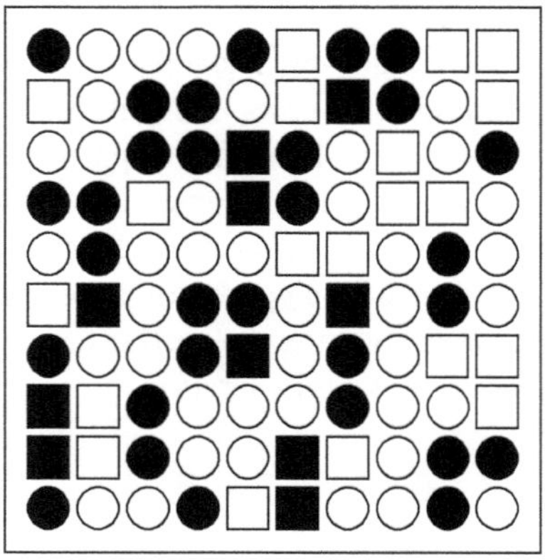

FIGURE 2.3 Typical example of patterns as a result of simulation. $r_1 = 1.1$, $r_2 = 2.1$, $k = 0$, I_Nov.

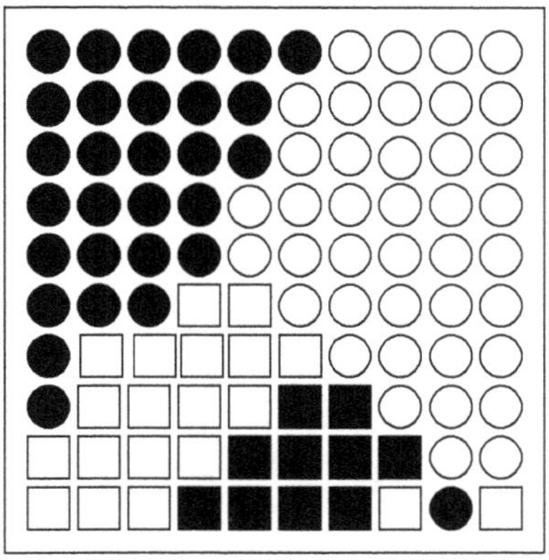

FIGURE 2.4 Typical example of patterns as a result of simulation. $r_1 = 1.5$, $k = 0$, I_0.

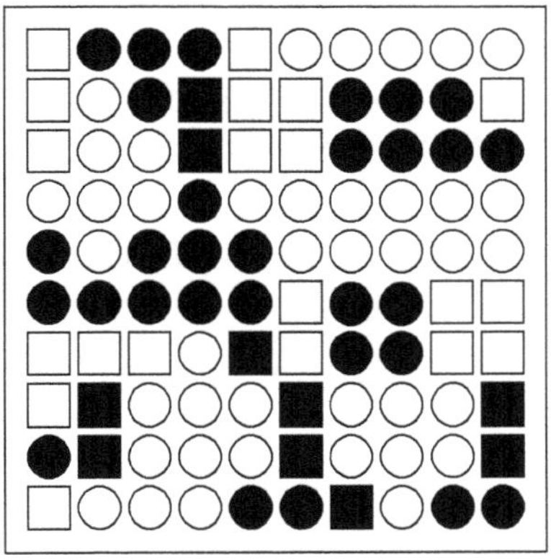

FIGURE 2.5 Typical example of patterns as a result of simulation. $r_1 = 1.5$, $r_2 = 2.9$, $k = 0$, I_N.

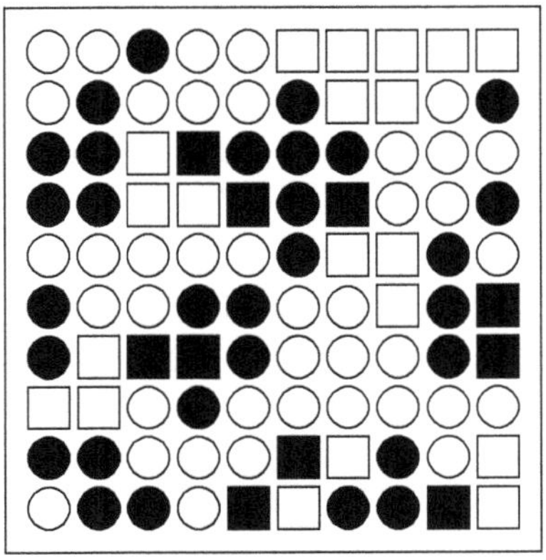

FIGURE 2.6 Typical example of patterns as a result of simulation. $r_1 = 1.5$, $r_2 = 2.9$, $k = 0$, I_Nov.

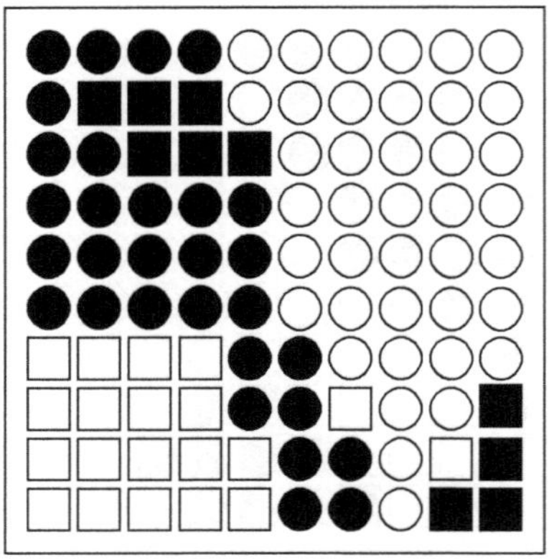

FIGURE 2.7 Typical example of patterns as a result of simulation. $r_1 = 1.1$, $k = 2$, I_0.

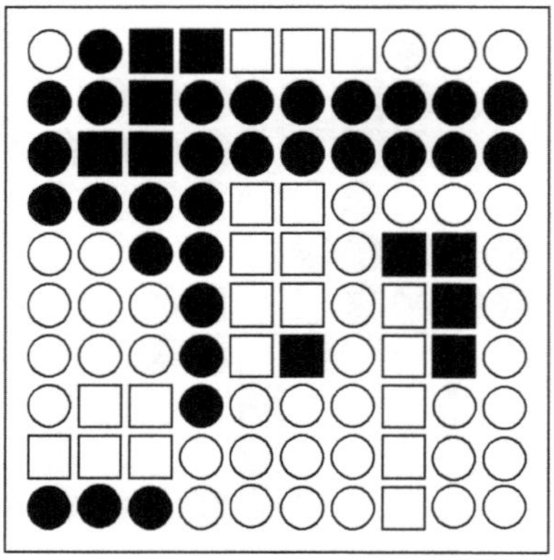

FIGURE 2.8 Typical example of patterns as a result of simulation. $r_1 = 1.1$, $r_2 = 2.1$, $k = 2$, I_N.

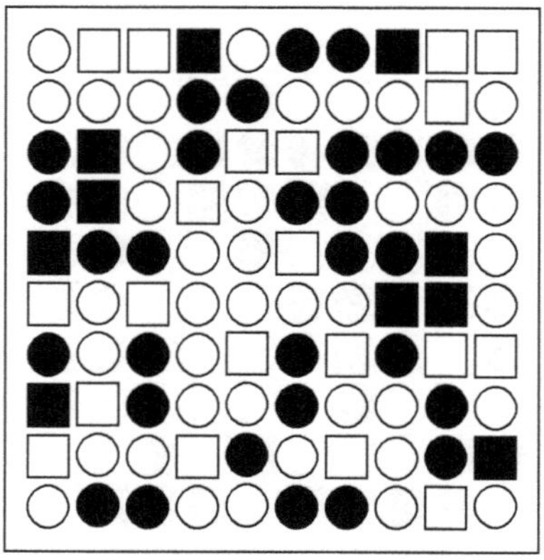

FIGURE 2.9 Typical example of patterns as a result of simulation. $r_1 = 1.1$, $r_2 = 2.1$, $k = 2$, I_Nov.

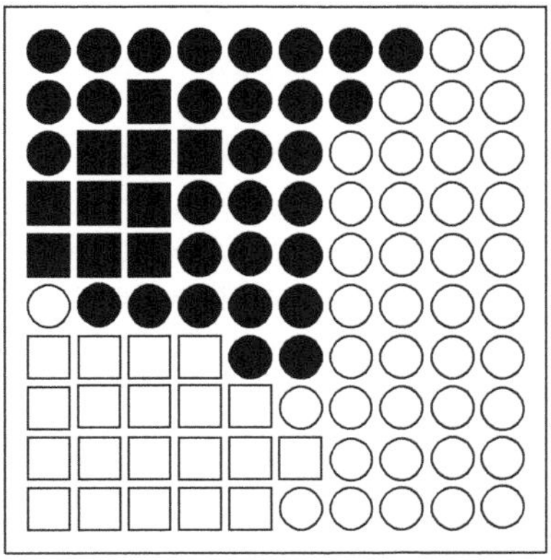

FIGURE 2.10 Typical example of patterns as a result of simulation. $r_1 = 1.5$, $k = 2$, I_0.

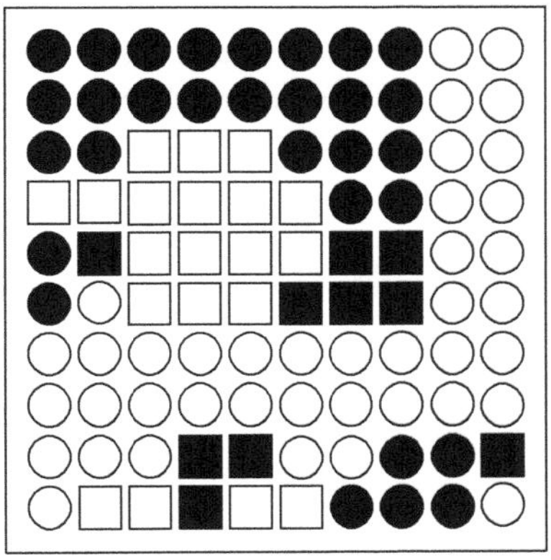

FIGURE 2.11 Typical example of patterns as a result of simulation. $r_1 = 1.5$, $r_2 = 2.9$, $k = 2$, I_N.

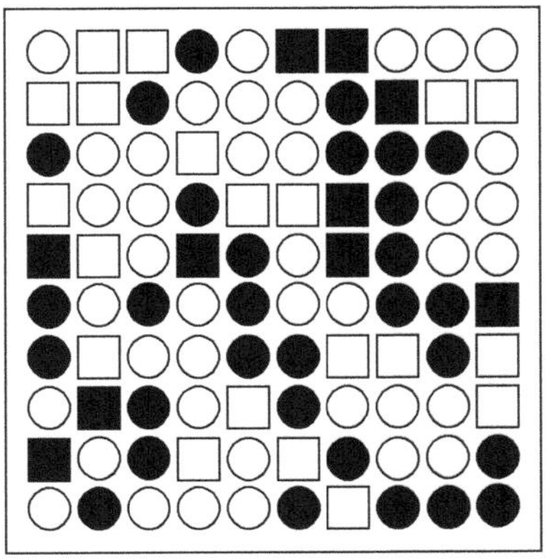

FIGURE 2.12 Typical example of patterns as a result of simulation. $r_1 = 1.5$, $r_2 = 2.9$, $k = 2$, I_Nov.

study refers to the condition that where the resistance to distance exists, the logics of purification and mixture are compounded and work together, but their levels are not overlaid but separated.

From the theoretical viewpoint of city planning, the coexistence of micro-purification and macro-mixture have been considered to be desirable in mixed-use districts where, for example, residential and industrial uses coexist—as pointed out by Okata (1994). Possible patterns of such mixed uses, however, have not yet been considered in detail. The simulation results of this study offer an insight into such patterns.

One distinctive example of a long agglomerative pattern is street-type development, as in the case of Makuhari Bay Town (Architectural Institute of Japan, 1997) in Chiba, in the suburbs of Tokyo. Some studies on residential planning (Omura, 1995; Inoue and Takada, 1998) have lauded it as a unique and successful project. Though not widely used, street-type development is gaining popularity in Japan. In this type of development, roads and avenues act as a catalyst, supporting and strengthening the development potential of uses.

Needless to say, there are numerous factors affecting city formation. This study's focus on the interaction between land use categories, however, should have some implications for more sensible and sustainable urban development.

2.5 CONCLUSIONS

This study simulating land use patterns was conducted to elicit implications for the theory of planning. Its two achievements were: (1) the development of a cell-based model for the simulation of land use patterns to demonstrate how cells influence each other by compounding two different levels of impact, and (2) the results of the simulation indicating an image resulting from the logic of micro-purification and macro-mixture.

Jacobs (1961) pointed out that scientific studies of cities need to treat them as "organized complexity." There are two opposite approaches to such studies, i.e., an empirical approach based on specific examples and evidence, and a theoretical approach based on a generalized model. This study is an attempt at the latter.

Many spatial logics have yet to be explored. With the advent of computers making it far easier to simulate complex interactions and systems, there will no doubt be greater possibilities for the development of statistical methods that will prove useful for the future analysis of cities and city planning.

ACKNOWLEDGMENTS

This study was conducted as part of a joint research project "Comprehensive Studies on Recycling-Oriented Society and City Planning" (FY1998–FY2001) by the then Tokyo Metropolitan University Center for Urban Studies (where the author was teaching) and the Center for Spatial Information Sciences of the University of Tokyo (where the author served as a visiting professor). The author thanks the staff of the two institutions for their cooperation and the information they supplied.

REFERENCES

Architectural Institute of Japan. 1997. Makuhari Baytown–Patios 4. *1997 Sakuhin Senshu* (1997 Selected Architectural Designs of the Architectural Institute of Japan), 94–95 (in Japanese).

Gatrell, A. C. 1977. Complexity and redundancy in binary maps. *Geographical Analysis* 9:29–41.

Inoue, S., and Takada, M. 1998. The influence of the layout on the residents' activity in street-type multi unit housing. *Journal of Architecture, Planning and Environmental Engineering* 505:91–96 (in Japanese).

Jacobs, J. 1961. *The death and life of great American cities*. New York: Random House.

Koide, O. 1977. Application and testing of the degree of land-use mixing. *Papers of the Annual Conference of the City Planning Institute of Japan* 12:79–84 (in Japanese).

Okata, J. 1994. Issues and perspectives in land-use planning. *Shin-toshi* 48:48–58 (in Japanese).

Okudaira, K. 1976. *Toshi Kogaku Dokuhon* (Urban engineering guidebook). Tokyo: Shokokusha.

Omura, K. 1995. Urban design and its issues: Makuhari new central district housing. *City Planning Review* 197:118–23 (in Japanese).

Portugali, J., Beneson, I., and Omer, I. 1994. Sociospatial residential dynamics: Stability and instability within a self-organizing city. *Geographical Analysis* 26:321–40.

Sugita, O., and Koshizuka, T. 1983. On a building quasi-coverage. *Papers of the Annual Conference of the City Planning Institute of Japan* 18:37–42 (in Japanese).

Tamagawa, H. 1982. The study of mathematical expressions on the order of land use. *Papers of the Annual Conference of the City Planning Institute of Japan* 12:73–78 (in Japanese).

Tamagawa, H. 2000. A compound simulation model of land use pattern and its implication. *Papers in City Planning* 35:1039–44 (in Japanese).

Yoshikawa, T. 1997. On a method to analyze the characteristics of land use agglomeration using grid data. *Journal of Architecture, Planning and Environmental Engineering* 495:147–54 (in Japanese).

Yoshikawa, T. 1998. An analysis of agglomeration of land use using same edge ratio in southern Tama district of Tokyo metropolis. *Comprehensive Urban Studies* 66:69–77 (in Japanese).

Yoshikawa, T. 1999. A method to analyze characteristics of agglomeration in the same and different kinds of land use using grid data. *Journal of Architecture, Planning and Environmental Engineering* 520:227–32 (in Japanese).

3 Optimal Hierarchical Transportation System with Economies of Scale*

Tsutomu Suzuki and Daisuke Watanabe

3.1 INTRODUCTION

In transportation of people or goods, carrying them over the shortest possible distance enables shorter transportation time and lower transportation cost. Typically, the cost and energy required for transportation increase with distance. At the same time, transporting a particular item in a certain quantity often results in labor savings. Mass transportation systems can achieve significant savings in energy and cost of transportation. If a certain quantity of people or goods is carried at one time, the total sum of cost and energy can be reduced, even with a certain number of transportation detours.

To reduce the total cost, we can define a hierarchy in the transportation system. For example, as illustrated in Figure 3.1, we assume the case of a demand for transportation to a particular point from several points located within a specified area. As shown in the diagram on the left, direct transportation to the destination point involves transportation from multiple points to just one, which is a laborious process. Thus, as shown in the diagram on the right, we can envision a method in which items are temporarily collected at a number of relay facilities to be carried to the destination point. This latter case results in a higher level of transportation of greater concentration. Then, as we get closer to the destination point, items are distributed from relay points to lower levels of the hierarchy, before ultimately arriving at the destination point. In this way, a transportation hierarchy is formed. We can refer to this as hierarchical transportation.

The question then arises: Under what conditions is this type of hierarchical transportation desirable?

Ogura (1999) set up a transportation network with a hierarchical structure to develop a door-to-door parcel delivery service. He noted that such a hierarchical system offers economies of scale; i.e., the transportation cost per unit of distance and weight gradually decreases as the transportation distance and weight increase. When there are economies of scale in transportation cost, it is advantageous to transport items in large quantities or over longer distances at one time, thereby resulting in hierarchical transportation.

* This was originally studied in Watanabe and Suzuki (2000) and is a summary chapter of a part of Watanabe's PhD thesis (2006) supervised by Suzuki.

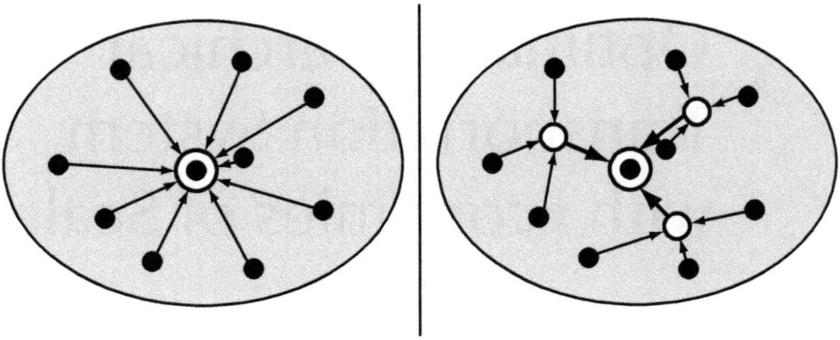

FIGURE 3.1 Hierarchical transportation.

In this chapter, we refer to the facilities that are viewed as nodes in hierarchical transportation as *transportation relay facilities*, and to generalization of the time and cost of transportation as *transportation cost*. By focusing on hierarchical transportation systems composed of multiple levels, we aim to determine optimal hierarchical structures and to investigate their relationship with the economies of scale in transportation cost.

In hierarchical transportation systems, items are transshipped to transportation relay facilities and then collected at higher-level facilities (or distributed to lower-level facilities) before being transported to the destination point. In this chapter, we determine the optimal number of levels of transportation hierarchies and the optimal number of facilities per level for minimizing the total transportation cost in the case where there are economies of scale in transportation cost. By assuming the number of facilities and number of hierarchy levels to behave approximately as continuous variables, we aim at overall optimization of large-scale transportation systems, even if our optimization is only approximate. Furthermore, we examine the impact of varying parameters that express economies of scale and increasing demand density, and study cases featuring round trip–type route patterns between facilities.

3.2 LITERATURE REVIEW

Central place theory is a classical mathematical model describing hierarchical structures. Christaller (1933) demonstrated that when it is assumed that a central place (city) is located on a triangular lattice, a strictly hierarchical structure and spatial patterns are formed, and the number of lower-level central places increases by a factor of k, depending on differences in the distances goods can reach. In addition, Lösch (1940) proposed a model in which k is generalized, deriving a hierarchical structure with a continuous scale distribution in the case of delivering multiple goods.

In the field of geomorphology, Horton's law, which describes the shape of stream branching, was verified in Horton (1945). According to Horton's first law, when hierarchical orders are assigned to river systems, the number of streams of order o is N_o, which is given by the relationship below:

$$N_o = r_b^{(S-o)} \tag{3.1}$$

Here, S is the highest stream order and r_b is the bifurcation ratio. Horton (1945) reported that this fits the behavior of real river systems. Shreve (1966)—in relation to random stream networks—and Tokunaga (1966)—in relation to the number of streams flowing into higher-order streams unrelated to order conversion—both confirmed that the bifurcation ratio decreases as the order becomes higher. With respect to traffic systems, Hagget (1967) conducted analyses on tree-structured expressways.

Problems concerning hierarchical structures are also studied in the field of facility location optimization. The size density hypothesis proposed by Stephan (1988) described the relationship between a population and number of facilities by applying the number of facilities as a continuous variable. This hypothesis shows that, in the case in which both the population and number of facilities are sufficiently large, optimization to minimize the transportation time with formulation of a continuous approximation model results in the general rule that the service area covered by one facility is inversely proportional to the demand density to the power of two thirds. Similarly, Kurita (1999) analytically determined the number of facilities for minimizing the total of transportation and facility costs in the case in which the facility costs can be expressed as a nonlinear function. In addition, Suzuki (2000) discussed the effectiveness of the size density hypothesis derived by continuous approximation of the number of facilities through a comparison with solutions to median problems.

Dökmeci (1973) examined the hierarchical structure of facilities for minimizing total cost—i.e., the sum of facility and transportation costs. Assuming a certain facility cost for each level in a system featuring facilities at three levels, he calculated the optimal number of levels and the number of facilities. He described a trade-off relationship between the two kinds of costs: as the number of facilities increases, transportation costs decrease, but facility costs increase.

In relation to models for optimizing the number of facilities in terms of the users' travel distance in user drop-off-type facilities, Suzuki (1990) formulated a location problem in continuous space, for which Okunuki and Okabe (1995) proposed an exploratory algorithm-based solution. Sadahiro and Okabe (1994) analyzed the hierarchical structure of the distribution of stores for real commercial facilities under conditions where lower-level facilities are allocated to the nearest higher-level facility.

Provided services have various characteristics according to the type of service, and various relationships can be observed between facilities providing higher-level services and facilities providing lower-level services. Narula (1984) conducted a unified classification in formulating facility location problems in these kinds of hierarchical systems, and proposed a classification using a service provision method based on facility hierarchy (categorized as successively inclusive or successively exclusive) and a transportation format based on graph theory (categorized as integrated or discriminating in terms of arc flow, and multipath or unipath in terms of node flow). Okabe and Okunuki (1993) reviewed hierarchical location problems using this classification. In addition, Church and Eaton (1987) presented a classification based on a service provision format (categorized as parallel service, sequential service, or referral service). Figure 3.2 (based on Ieda, 1997) summarizes these classifications. The transportation covered in this chapter is shown at the bottom right of Figure 3.2.

In the optimization of transportation systems, various types of models have been proposed in the field of operations research, such as models for facility location

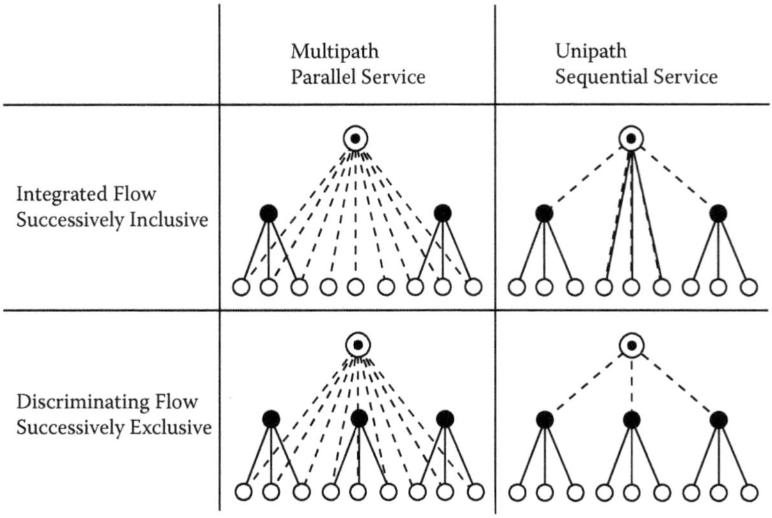

FIGURE 3.2 Classification of hierarchical systems.

problems and vehicle routing problems. Eilon et al. (1971) summarized some basic models for facility location and delivery routing for the purpose of applying them to distribution systems. In recent years many models have been developed to handle sophisticated distribution systems—for supply chain management, for example—and these have been comprehensively reviewed by Crainic and Laporte (1997). Detailed reviews of facility location in distribution systems have also been conducted by Daskin (1985) and by Klose and Drexl (2005), among others. Cost functions for transportation costs have been classified by Higginson (1993).

In the area of facility location problems a lot of research has been done on modeling using continuous approximation to overcome the problem that discrete optimization cannot be used to solve large-scale problems. A detailed review of this work was conducted by Langevin et al. (1996). Daganzo (1999) formulated a number of models relating to facility and operation intervals and inventory, and performed analytical calculations for optimizing distribution systems. Studies utilizing continuous approximation for facility location include those of Campbell (1990), who first examined cases where there are economies of scale in transportation costs relating to transportation flow, and then investigated routing (Campbell, 1993). Suzuki and Kawaguchi (1998) examined an optimal transshipment facility location in a one-dimensional city, and Kasahara and Furuyama (1998) investigated the number of area divisions for minimizing total length of tree.

Most studies dealing with the verification of hierarchical structures of Japanese distribution systems have been conducted in the field of civil engineering planning studies. To optimize vehicle stop intervals, Matsumoto (1990) developed a model for the trade-off between transportation and inventory costs, and Ieda et al. (1992) developed a model that takes into account transverse travel. Tokunaga et al. (1995) constructed a model for simultaneously determining collection/delivery center locations and collection/delivery routes. Ishiguro et al. (2000) constructed a model for minimizing total transportation costs based on an incremental assignment method and taking into account economies of scale relating to transportation volume.

Micro-level studies focused on transportation volume-related economies of scale and aimed at deriving the optimal number of transportation hierarchy levels, including those of Ieda (1997) and Kijmanawat and Ieda (2004), who developed hub-and-spoke-type models of the facility location problem, based on dynamic and mathematical programming, respectively.

Location problems for transportation relay facilities in which the number of hierarchy levels is fixed at two have been studied, but there have been virtually no studies aimed at optimizing the number of transportation hierarchy levels. The works of Ieda et al. are unusual in that they focus on this issue, but they only investigated the characteristics of parameters based on numerical computation, without touching on economies of scale in relation to transportation distance. In this chatper we focus on precisely this issue, but since it is generally difficult to determine globally optimal solutions when the number of demand points and facilities is large, we have developed a model that enables continuous approximation.

3.3 HIERARCHICAL TRANSPORTATION MODEL

In this chapter, as shown in Figure 3.3, we constructed a hierarchical transportation system model that assumes that the goods needing to be transported and distributed within a certain area are collected at several places. We made the following assumptions regarding the transportation demand within the area and the facilities through which goods pass:

1. A total of n_0 facilities are evenly distributed at the lowest level of the hierarchy, and there is an evenly distributed transportation demand to carry the goods to n_M facilities at the highest level of the hierarchy. The quantity

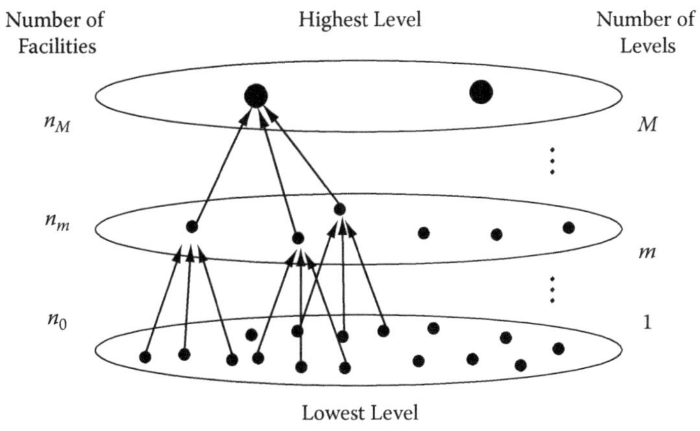

FIGURE 3.3 Hierarchical transportation model.

of demand assigned to one facility at the highest level is n_0/n_M. Note that demand does not increase at intermediate hierarchy levels, and the relationship between facilities at different levels of the hierarchy is such that the service is regarded as successively exclusive.

2. In terms of collection/delivery routes, the goods arrive at the highest level after passing through M levels of hierarchical transportation, as shown in Figure 3.3. Travel between facilities is nonstop and straight-line. This is referred to as direct-type or single-stop (SS) transportation.
3. We assume that n_m facilities are uniformly allocated to m-th level intermediate facilities ($1 \leq m \leq M$) within the service area. These facilities serve as destination points of the m-th level transportation, where the demand collected up to the (m – 1)-th level is delivered to the nearest m-th level facility. The aggregated demand is then sent on to the (m + 1)-th level of transportation.
4. The number of facilities at each level satisfies $n_0 \ll n_1 \ll \ldots \ll n_m \ll \ldots \ll n_M$. The number of levels m and number of facilities n_m are strictly natural numbers, but on the assumption that they are sufficiently large, we consider them suitable for continuous approximation.

Note that in the case of delivery, the process is completely the opposite to that of transportation, so the same formulation can be used.

Generally with transportation costs, we can safely assume that the incremental change in transportation cost—i.e., marginal cost—gradually decreases as transportation volume and transportation distance increase. Thus, we assume that the transportation cost is determined by transportation volume and transportation distance, such that the transportation cost C when carrying goods of transportation volume q over a distance l can be expressed by the following power function relating transportation volume to transportation distance, where K is a coefficient representing fixed costs:

$$C = Kq^{\alpha}l^{\beta} \qquad (0 \leq \alpha \leq 1, 0 < \beta \leq 1) \tag{3.2}$$

Here, α is the transportation volume elasticity relative to transportation cost, and β is the transportation distance elasticity. Figure 3.4 shows a typical model for transportation cost C for varying transportation volume q and transportation distance l. It is clear that, as the transportation volume and distance increase, the marginal rise in transportation cost gradually decreases. In this chapter we refer to this feature as economies of scale. The greater the economies of scale, the lower this marginal cost becomes. The value of α depends on the characteristics of the transported goods, while β also (in addition to goods) depends on long-distance, high-speed transportation technology.

Next, we consider the total transportation cost over a particular service area. This varies according to the shape of the area, but we can make a general formulation for each spatial dimension.

First, we consider the case in which the service area is a one-dimensional linear city of total length L. The collection/delivery is conducted as shown in Figure 3.5. If we assume that goods are transported from n_{m-1} facilities at the (m – 1)-th level to n_m

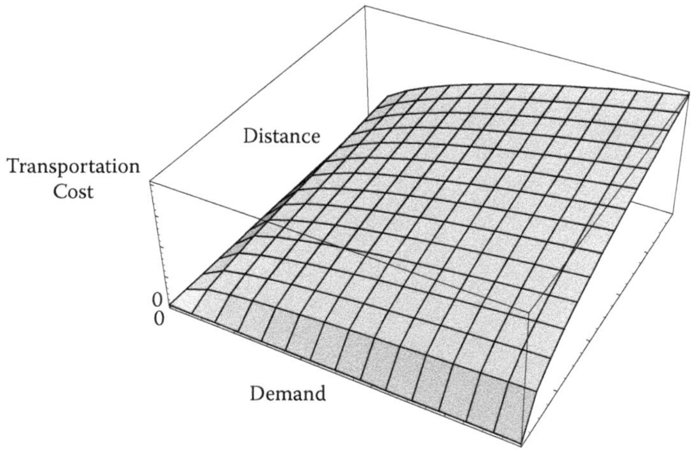

FIGURE 3.4 Relationship between demand, transportation distance, and transportation cost.

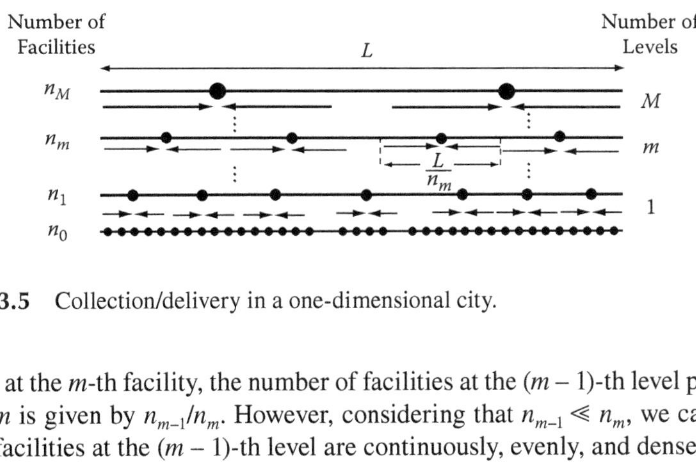

FIGURE 3.5 Collection/delivery in a one-dimensional city.

facilities at the m-th facility, the number of facilities at the $(m-1)$-th level per facility at level m is given by n_{m-1}/n_m. However, considering that $n_{m-1} \ll n_m$, we can assume that the facilities at the $(m-1)$-th level are continuously, evenly, and densely distributed over the length L/n_m, which is the area of collection of each facility. Since transportation volume $q_m = n_0/n_{m-1}$ is transported over a distance l from n_{m-1}/L facilities at the $(m-1)$-th level toward the center of the collection area, the transportation cost at the m-th level, taking into account economies of scale, can be expressed as follows:

$$c_m^l = n_m 2\int_0^{\frac{L}{2n_m}} \frac{n_{m-1}}{L}\left(\frac{n_0}{n_{m-1}}\right)^{\alpha} l^{\beta}\,dl = \frac{n_0^{\alpha}}{\beta+1}\left(\frac{L}{2}\right)^{\beta}\frac{n_{m-1}^{1-\alpha}}{n_m^{\beta}} \tag{3.3}$$

When the area is two-dimensional, various area shapes are possible. Here, we consider that collection/delivery is carried out toward n_m facilities at the m-th level of the transportation hierarchy in a circular city of area S (Figure 3.6). The radius of the circular area at the m-th level is given by $\sqrt{S/{\neq}n_m}$. Since transportation volume $q_m = n_0/n_{m-1}$ is transported over a distance l from n_{m-1}/S facilities distributed continuously

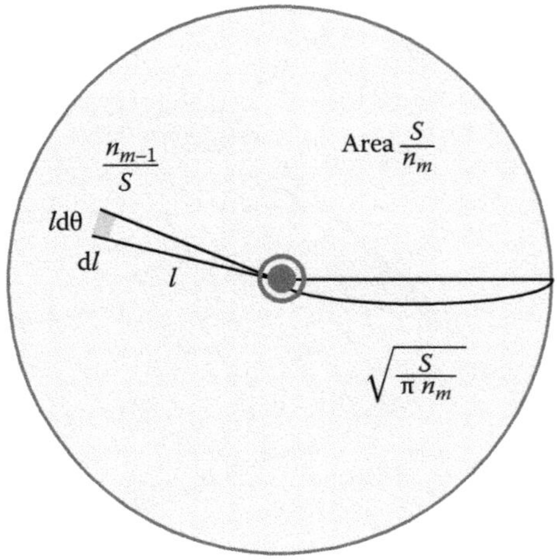

FIGURE 3.6 Transportation to the m-th level in a circular city.

within the area toward the center of the circle, the transportation cost at the m-th level, taking into account economies of scale, can be expressed as follows:

$$c_m^c = n_m \int_0^{2\pi} \int_0^{\sqrt{S/\pi n_m}} \left(\frac{n_{m-1}}{S} \frac{n_0}{n_{m-1}} \right)^{\alpha} l^{\beta} l dl d\theta = \frac{2n_0^{\alpha}}{\beta+2} \left(\frac{S}{\pi} \right)^{\frac{\beta}{2}} \frac{n_{m-1}^{1-\alpha}}{n_m^{\beta/2}} \tag{3.4}$$

If we consider all the items unrelated to the number of levels m collectively as the transportation cost coefficient, we can make a general formulation to express total transportation cost, independent of the shape or number of dimensions of the service area, by substituting the economies of scale parameters α and β and the transportation cost coefficient with a, b, and K, as shown in Table 3.1.

$$c_m = K \frac{n_{m-1}^{a}}{n_m^{b}}, \qquad m = 1, \ldots, M \tag{3.5}$$

It should be noted here that the value range of a, b will vary according to the domains of α and β.

TABLE 3.1
Substitution of Parameters (SS)

c_m	K	a	b	Dimension
c_m^l	$K_l = n_0^{\alpha} \frac{1}{\beta+1} \left(\frac{L}{2} \right)^{\beta}$	$1-\alpha$	β	1-dimensional SS
c_m^c	$K_c = \frac{2n_0^{\alpha}}{\beta+2} \left(\frac{S}{\pi} \right)^{\frac{\beta}{2}}$	$1-\alpha$	$\frac{\beta}{2}$	2-dimensional SS

The hierarchical transportation system to be determined can be obtained by deriving the number of hierarchy levels M^* and number of m-th level facilities n_m^{**} that minimize the generalized total transportation cost C_t when the demand quantity n_0 and the number of facilities n_M at the highest hierarchy levels are given—that is, by solving the following minimization problem:

$$Min \quad C_t = \sum_{m=1}^{M} c_m = K \sum_{m=1}^{M} \frac{n_{m-1}^a}{n_m^b} \tag{3.6}$$

3.4 OPTIMAL HIERARCHICAL STRUCTURE BY MINIMIZATION OF TRANSPORTATION COST

The optimal hierarchical transportation system is obtained by deriving the number of hierarchy levels M^* and number of m-th level facilities n_m^{**} that minimize the total transportation cost (Equation (3.6)). For this it is necessary to categorize the problem into three different cases, according to the values of a and b:

1. $a = 0$: No economies of scale exist in relation to the transportation volume.
2. $a = b$: There is a fixed relationship between the economies of scale relating to transportation volume and transportation distance.
3. $a \neq b$: General case.

3.4.1 No Economies of Scale in Relation to Transportation Volume

Substituting $a = 0$ in Equation (3.6) results in the following expression for total transportation cost:

$$C_t = K \sum_{m=1}^{M} \frac{1}{n_m^b} \tag{3.7}$$

The first derivative of this expression with respect to the number of facilities can be approximated as follows, if n_m is sufficiently large:

$$\frac{\partial C_t}{\partial n_m} = -\frac{Kb}{n_m^{b+1}} < 0 \tag{3.8}$$

Accordingly, it is clear that the optimal transportation system in this case is direct transportation with $M^* = 1$, where there is no intermediate collection/transportation.

3.4.2 Fixed Relationship between the Economies of Scale Relating to Transportation Volume and Transportation Distance

In the case where $a = b$, we can substitute $a = b$ into the formula for total transportation cost (Equation (3.6)) to obtain

$$C_t = K\sum_{m=1}^{M}\left(\frac{n_{m-1}}{n_m}\right)^a \tag{3.9}$$

If we take the first-level conditions for the number of facilities n_m to minimize this expression of cost, we get

$$\frac{\partial C_t}{\partial n_m} = -\frac{Kan_{m-1}^a}{n_m^{a+1}} + \frac{Kan_m^{a-1}}{n_{m+1}^a} = 0 \tag{3.10}$$

This can be simplified to the following relation:

$$\frac{n_m}{n_{m+1}} = \frac{n_{m-1}}{n_m}, \qquad m = 1,\ldots,M-1 \tag{3.11}$$

This expression indicates that the ratio between the number of facilities at neighboring levels is always constant. In addition, since $\partial^2 C_t / \partial n_m^2 > 0$, the total transportation cost C_t is a convex downward function, indicating the existence of a minimum value. Expanding the relationship of Equation (3.11), we obtain

$$\frac{n_0}{n_m} = \frac{n_{m-1}}{n_m}\frac{n_{m-2}}{n_{m-1}}\cdots\frac{n_0}{n_1} = \left(\frac{n_0}{n_1}\right)^m \tag{3.12}$$

The optimal number of facilities at the m-th level, n_m*, can be expressed as shown below, using the given values n_0 and n_M:

$$n_m^* = n_0\left(\frac{n_M}{n_0}\right)^{\frac{m}{M}} \tag{3.13}$$

The total transportation cost C_t* when Equation (3.13) for the number of facilities is applicable can be expressed as follows:

$$C_t^* = K\sum_{m=1}^{M}\left(\frac{n_0}{n_M}\right)^{\frac{a}{M}} = KM\left(\frac{n_0}{n_M}\right)^{\frac{a}{M}} \tag{3.14}$$

Next, we calculate the number of hierarchy levels that minimize the total transportation cost of this hierarchically structured transportation system. If we assume first-level conditions for the number of levels M relating to Equation (3.6) expressing total transportation cost, we get

$$\frac{\partial C_t^*}{\partial M} = K\left(\frac{n_0}{n_M}\right)^{\frac{a}{M}}\left(1 - \frac{a}{M}\log\frac{n_0}{n_M}\right) = 0 \tag{3.15}$$

and the optimal number of levels M^* can thus be expressed as

$$M^* = a \log \frac{n_0}{n_M} \tag{3.16}$$

Since $\partial^2 C_t^* / \partial M^2 > 0$ is satisfied, Equation (3.16) defines the number of levels that result in the minimum total transportation cost. This means that the optimal number of levels corresponds to the product of parameter a, which indicates economies of scale, and the logarithm of the ratio of the number of facilities at the lowest level (demand) to the number of facilities at the highest level.

From Equation (3.13) the optimal number of facilities at the m-th level, n_m^{**}, can be expressed as

$$n_m^{**} = n_0 e^{-m/a} \tag{3.17}$$

The total transportation cost C^{**} for the optimal hierarchical transportation system can be determined by the formula below, based on the expression of Equation (3.14) for total transportation cost.

$$C_t^{**} = Kae \log \frac{n_0}{n_M} \tag{3.18}$$

In this case, the transportation cost c_m^{**} for transportation at the m-th level can be expressed by the expression below, based on Equation (3.5).

$$c_m^{**} = Ke \tag{3.19}$$

This indicates a fixed cost, independent of the hierarchy level m.

3.4.3 General Case

Although slightly more complicated than the case in Section 3.4.2, even in the case where $a \neq b$, the optimal transportation system can generally be determined in the way described above. The first-level conditions relating to number of facilities n_m in Equation (3.6) can be expressed as

$$\frac{\partial C_t}{\partial n_m} = -\frac{Kbn_{m-1}^a}{n_m^{b+1}} + \frac{Kan_m^{a-1}}{n_{m+1}^b} = 0 \tag{3.20}$$

This can be simplified to

$$\frac{n_m}{n_{m+1}} = \left(\frac{b}{a}\right)^{1/b} \left(\frac{n_{m-1}}{n_m}\right)^{a/b}, \qquad m = 1, \ldots, M-1 \tag{3.21}$$

Since $\partial^2 C_t / \partial n_m^2 > 0$, the total transportation cost C_t has a minimum value. Expanding Equation (3.21) we obtain

$$\frac{n_0}{n_m} = \frac{n_{m-1}}{n_m}\frac{n_{m-2}}{n_{m-1}}\cdots\frac{n_0}{n_1} = \left(\frac{b}{a}\right)^{\frac{1}{b-a}\left(m-\frac{1-(a/b)^m}{1-a/b}\right)}\left(\frac{n_0}{n_1}\right)^{\frac{1-(a/b)^m}{1-a/b}} \tag{3.22}$$

The optimal number of facilities at the m-th level, n_m*, can be expressed as shown below, using the given values n_0 and n_M.

$$n_m^* = n_0\left(\frac{b}{a}\right)^{\frac{1}{a-b}\left(m-\frac{1-(a/b)^m}{1-(a/b)^M}M\right)}\left(\frac{n_M}{n_0}\right)^{\frac{1-(a/b)^m}{1-(a/b)^M}} \tag{3.23}$$

The total transportation cost C_t* when Equation (3.23) for the number of facilities is applicable can be expressed as

$$C_t^* = Kn_0^{a-b}\left(\frac{a}{b}\right)^{\frac{b}{a-b}+\frac{M}{1-(a/b)^M}}\frac{a}{a-b}\left(1-\left(\frac{b}{a}\right)^M\right)\left(\frac{n_0}{n_M}\right)^{\frac{b-a}{1-(a/b)^M}} \tag{3.24}$$

As in Section 3.4.2, in this hierarchical structure we calculate the number of levels that minimize the total transportation cost of this hierarchically structured transportation system. If we assume first-level conditions for the number of levels M relating to Equation (3.14) that express total transportation cost, we obtain

$$\begin{aligned}\frac{\partial C_t^*}{\partial M} = {} & Kn_0^{a-b}\frac{b}{a-b}\left(\frac{a}{b}\right)^{\frac{a}{a-b}+\frac{M}{1-(a/b)^M}}\left(\frac{n_0}{n_M}\right)^{\frac{b-a}{1-(a/b)^M}} \\ & \frac{1}{(a/b)^M-1}\log\frac{a}{b}\left(M\log\frac{a}{b}+(b-a)\log\frac{n_0}{n_M}\right) = 0\end{aligned} \tag{3.25}$$

Accordingly, the optimal number of levels M* can be expressed as

$$M^* = \frac{a-b}{\log(a/b)}\log\frac{n_0}{n_M} \tag{3.26}$$

Since $\partial^2 C_t^* / \partial M^2 > 0$ is satisfied, Equation (3.26) defines the number of levels that result in the minimum total transportation cost. When a and b are infinitely close in value, it can be proved that this expression is equivalent to Equation (3.16).

If Equation (3.26) for the optimal number of levels is modified, the optimal number of facilities at the m-th level, n_m**, is shown as

$$n_m^{**} = n_0 \left(\frac{a}{b}\right)^{\frac{m}{b-a}} \tag{3.27}$$

By substituting into Equation (3.24) for total transportation cost, the total transportation cost C_t** with an optimal number of hierarchy levels can be determined as

$$C_t^{**} = K n_0^{a-b} \frac{b}{b-a} \left(\frac{a}{b}\right)^{\frac{a}{a-b}} \left[\left(\frac{n_0}{n_M}\right)^{b-a} - 1\right] \tag{3.28}$$

The transportation cost at the m-th level, c_m**, can be calculated as follows, based on Equation (3.5):

$$c_m^{**} = K n_0^{a-b} \left(\frac{a}{b}\right)^{\frac{a}{a-b}-m} \tag{3.29}$$

To express the above results using the parameters α and β as shown in Table 3.1, which represent economies of scale for each dimension of the service area, the following equations can be derived:

One-dimensional SS:

$$M^* = \begin{cases} \beta \log \dfrac{n_0}{n_M} & \textit{if} \quad \alpha + \beta = 1, \\ \dfrac{1-\alpha-\beta}{\log\{(1-\alpha)/\beta\}} \log \dfrac{n_0}{n_M} & \text{otherwise} \end{cases} \tag{3.30}$$

$$n_m^{**} = \begin{cases} n_0 e^{-m/\beta} & \textit{if} \quad \alpha + \beta = 1, \\ n_0 \left(\dfrac{1-\alpha}{\beta}\right)^{\frac{m}{\alpha+\beta-1}} & \text{otherwise} \end{cases} \tag{3.31}$$

Taking the logarithm of Equation (3.31), we find the following expression:

$$\log n_m^{**} = \begin{cases} \log n_0 - \dfrac{m}{\beta} & \textit{if} \quad \alpha + \beta = 1, \\ \log n_0 - \dfrac{m}{1-\alpha-\beta} \log \dfrac{1-\alpha}{\beta} & \text{otherwise} \end{cases} \tag{3.32}$$

This shows a first-order decreasing function of the number of hierarchy levels m.

Two-dimensional SS:

$$M^* = \begin{cases} \dfrac{\beta}{2}\log\dfrac{n_0}{n_M} & \text{if} \quad \alpha + \dfrac{\beta}{2} = 1, \\ \dfrac{1-\alpha-\beta/2}{\log\{2(1-\alpha)/\beta\}}\log\dfrac{n_0}{n_M} & \text{otherwise} \end{cases} \tag{3.33}$$

$$n_m^{**} = \begin{cases} n_0 e^{-2m/\beta} & \text{if} \quad \alpha + \dfrac{\beta}{2} = 1, \\ n_0 \left(\dfrac{1-\alpha}{\beta/2}\right)^{\frac{m}{\alpha+\beta/2-1}} & \text{otherwise} \end{cases} \tag{3.34}$$

When $a = b$, i.e., in the case where $\alpha + \beta/2 = 1$, the domains of α and β are defined as $1/2 \le \alpha < 1$ and $0 < \beta \le 1$, respectively. The logarithm of Equation (3.34) results in the following expression:

$$\log n_m^{**} = \begin{cases} \log n_0 - \dfrac{2m}{\beta} & \text{if} \quad \alpha + \dfrac{\beta}{2} = 1, \\ \log n_0 - \dfrac{m}{1-\alpha-\beta/2}\log\dfrac{1-\alpha}{\beta/2} & \text{otherwise} \end{cases} \tag{3.35}$$

Each of these cases represents first-order decreasing functions of the number of hierarchy levels m.

The results for both one-dimensional SS and two-dimensional SS show that, as the transportation volume elasticity α decreases and the economies of scale in relation to transportation volume increase, the optimal number of transportation levels increases, indicating that more frequent transshipment is desirable. In addition, as the transportation distance elasticity β decreases and the economies of scale in relation to transportation distance increase, the optimal number of transportation levels decreases, indicating that it is best to avoid transshipment as far as possible (Figures 3.7 and 3.8). It is clear that the optimal number of hierarchy levels is undoubtedly higher in the case of one-dimensional SS than in the case of two-dimensional SS. Furthermore, as the density of the transportation demand increases, the optimal number of levels increases, but at most this increase occurs on a logarithmic scale relative to demand growth.

3.5 DERIVATION OF OPTIMAL TRANSPORTATION SYSTEM IN RELATION TO ROUND-TRIP ROUTES

Up to now, we have developed models of hierarchical transportation systems for the case in which economies of scale exist in relation to transportation cost by assuming the number of facilities and number of hierarchy levels to be continuous

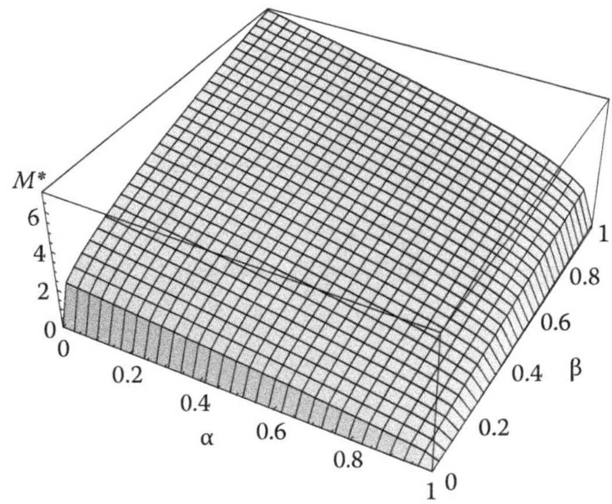

FIGURE 3.7 Relationship between economies of scale, α and β, and optimal number of hierarchy levels M^* (one-dimensional SS).

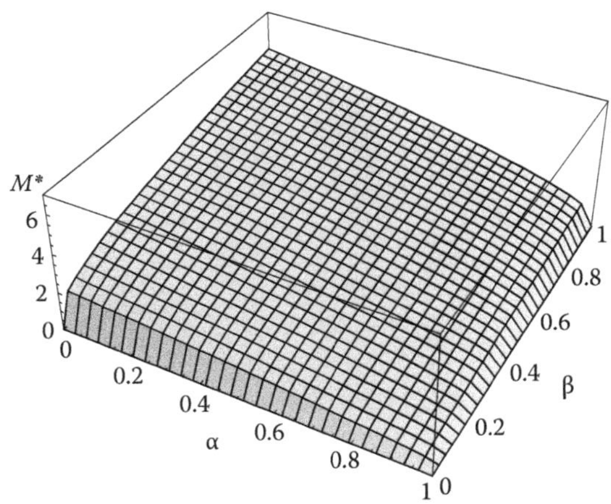

FIGURE 3.8 Relationship between economies of scale, α and β, and optimal number of hierarchy levels M^* (two-dimensional SS).

quantities. Our models are focused on single-stop (SS) transportation, i.e., without stops between facilities. In real transportation systems, however, vehicles typically stop at multiple places on their routes. We therefore defined the following typical transportation route patterns (Figure 3.9):

1. Direct transportation: single stop (SS), multi-vehicle (MV)
2. Delivery-type transportation: Multi-stop (MS), multi-vehicle (MV)
3. Round-trip type: Multi-stop (MS), single vehicle (SV)

Delivery-type transportation is difficult to formulate, because there are both multiple stops and multiple vehicles. Conversely, in round trip–type transportation, a

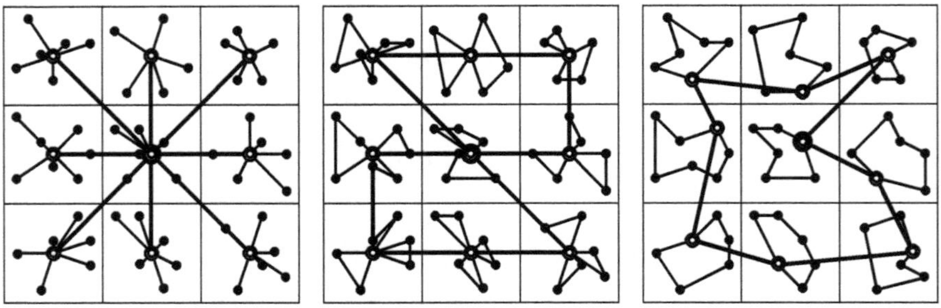

FIGURE 3.9 Transportation route selection methods.

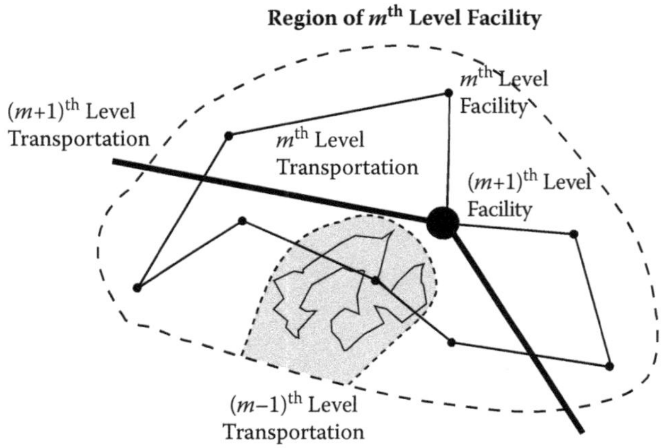

FIGURE 3.10 Collection/delivery in a two-dimensional city (MS).

single vehicle covers all the demand points in an allocated area, making formulation relatively easy if route distances are determined. Delivery-type transportation, which is intermediate between direct-type and round trip–type transportation, can also be considered to be intermediate in terms of the number of hierarchy levels determined for the other two methods.

Therefore, in this section, we change the travel route pattern (hypothesis 2 in Section 3.3 above), as described below, and develop a model for round trip–type transportation with a multiple-level hierarchy in a two-dimensional space (multi-stop (MS)). We also examine the impact of parameters representing economies of scale:

> 2′ (as changed from Section 3.3, hypothesis 2). The goods arrive at the highest level of the hierarchy after passing through M levels of transportation, as shown in Figure 3.10. Travel between facilities is assumed to be round trip–type transportation, with multiple stops along the route.

Using the BHH theorem derived by Beardwood et al. (1959), we determine an approximate value for the length of the round-trip route. It is known that in a two-dimensional area A, the expected value l of the shortest distance to cover n random points can be expressed as follows, where k represents a constant of proportionality:

$$l = k\sqrt{nA} \tag{3.36}$$

Thus, the transportation distance l_m from facilities at the m-th level can be expressed as follows by substituting $A = S/n_m$, the area covered by these facilities and the number of facilities at the $(m-1)$-th level, $n = n_{m-1}/n_m$, into Equation (3.36):

$$l_m = \frac{k\sqrt{n_{m-1}S}}{n_m} \tag{3.37}$$

The transportation demand at the m-th level of transportation is assumed to be equal to the load capacity of the truck, which is considered equivalent to the total transportation demand within the coverage area, expressed as

$$q_m = \frac{n_0}{n_m} \tag{3.38}$$

The transportation cost at the m-th level can be calculated as a product of the transportation volume q_m and transportation distance l_m, taking into account economies of scale, multiplied by the number of facilities n_m, as shown:

$$c_m^p = n_m q_m^\alpha l_m^\beta = n_0^\alpha k^\beta S^{\beta/2} \frac{n_{m-1}^{\beta/2}}{n_m^{\alpha+\beta-1}} \tag{3.39}$$

Substituting the economies of scale parameters α and β, and the transportation cost coefficient in Equation (3.39) with a, b, and K, respectively, as shown in Table 3.2, results in a generalized formula for transportation cost, independent of area shape. In this way, Equation (3.39) for the transportation cost at each hierarchy level results in the same shape as Equation (3.5). Accordingly, since this will be the same as Equation (3.6) for total transportation cost, the result derived earlier can be used. The domains for α and β, however, will be defined by $0 \le \alpha \le 1$, $0 < \beta \le 1$, and $\alpha + \beta > 1$, respectively, given that $a > 0$ and $b > 0$.

TABLE 3.2
Substitution of Parameters (MS)

c_m	K	a	b	Dimension
c_m^p	$K_p = n_0{}^\alpha k^\beta S^{\frac{\beta}{2}}$	$\frac{\beta}{2}$	$\alpha + \beta - 1$	2-dimensional MS

The optimal number of hierarchy levels and facilities in round trip–type transportation can both be derived as before. The results are expressed as

$$M^* = \begin{cases} \dfrac{\beta}{2}\log\dfrac{n_0}{n_M} & \textit{if} \quad \alpha + \dfrac{\beta}{2} = 1, \\ \dfrac{\alpha + \beta/2 - 1}{\log\{2(\alpha+\beta-1)/\beta\}}\log\dfrac{n_0}{n_M} & \text{otherwise} \end{cases} \tag{3.40}$$

$$n_m^{**} = \begin{cases} n_0 e^{-2m/\beta} & \textit{if} \quad \alpha + \dfrac{\beta}{2} = 1, \\ n_0 \left(\dfrac{\beta/2}{\alpha+\beta-1}\right)^{\frac{m}{\alpha+\beta/2-1}} & \text{otherwise} \end{cases} \tag{3.41}$$

Taking the logarithm of Equation (3.41) results in

$$\log n_m^{**} = \begin{cases} \log n_0 - \dfrac{2m}{\beta} & \textit{if} \quad \alpha + \dfrac{\beta}{2} = 1, \\ \log n_0 - \dfrac{m}{1-\alpha-\beta/2}\log\dfrac{\beta/2}{\alpha+\beta-1} & \text{otherwise} \end{cases} \tag{3.42}$$

Each of these cases represents a first-order decreasing function of the number of hierarchy levels m. As the transportation distance elasticity β decreases, the number of hierarchy levels declines rapidly. Conversely, as transportation volume elasticity α decreases, the number of hierarchy levels decreases, in contrast with the case of two-dimensional SS (Figure 3.11).

3.6 CONCLUSION

In this chapter, we studied the transportation costs in a hierarchically structured transportation system to derive the number of hierarchy levels and number of facilities at each level to minimize the transportation cost by taking into account economies of scale—i.e., the fact that the marginal increase in transportation cost gradually falls as transportation volume and transportation distance increase. In addition, we examined the changes in optimal transportation systems owing to parameters such as economies of scale and demand. The results of these evaluations clarified the following three points.

First, in the case of direct-type (single-stop (SS)) collection/delivery routes, the optimal number of transportation levels increases as the economies of scale in relation to transportation volume increase, but the optimal number of transportation levels decreases as the economies of scale in relation to transportation distance increase. For the same demand, a one-dimensional city requires a higher number of hierarchy levels than a two-dimensional city.

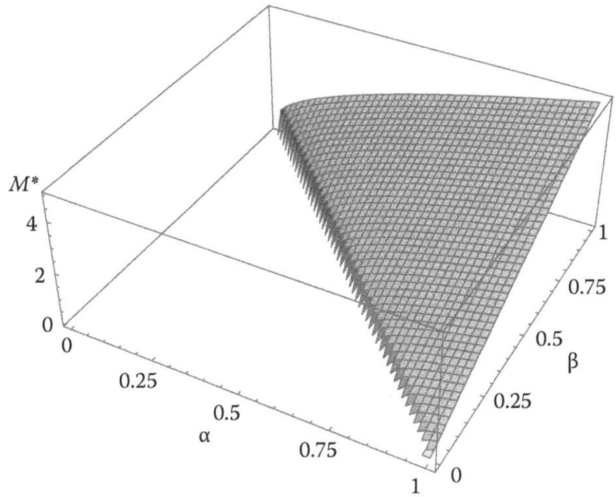

FIGURE 3.11 Relationship between economies of scale, α and β, and optimal number of hierarchy levels M^* (two-dimensional MS).

Second, we found that as the number of hierarchy levels increases, the optimal number of facilities at intermediate hierarchy levels increases in proportion to the logarithm of the number of facilities. When demand increases, the number of hierarchy levels increases, but at most the increase is proportional to the logarithm of the ratio of the number of facilities at the lowest level (demand) to the number of facilities at the highest level.

Third, in the case of a round trip–type collection/delivery route (MS), the higher the economies of scale in relation to transportation volume and transportation distance, the lower the number of transportation levels. Comparing this with the case of single-stop (SS) collection/delivery, we find that the results for economies of scale in relation to transportation volume are the converse.

In this chapter we proposed strong hypotheses of collection/delivery routes and demand in order to focus on the relationship between economies of scale and transportation costs and number of levels in the transportation hierarchy. A further challenge, as we look ahead, is to derive optimal transportation systems for more generalized models.

REFERENCES

Beardwood, J., Halton, J. H., and Hammersley, J. M. 1959. The shortest path through many points. *Proceedings of the Cambridge Philosophical Society* 55:299–327.

Campbell, J. F. 1990. Freight consolidation and routing with transportation economies of scale. *Transportation Research B* 24B:345–61.

Campbell, J. F. 1993. One-to-many distribution with transshipments: An analytic model. *Transportation Science* 27:330–40.

Christaller, W. 1933. *Die zentralen Orte in Süddeutschland*. Jena, Germany: Fischer.

Church, R. L., and Eaton, D. J. 1987. Hierarchical location analysis using covering objectives. In *Spatial analysis and location-allocation models*. New York: Van Nostrand Reinhold Company, 163–85.

Crainic, T. G., and Laporte, G. 1997. Planning models for freight transportation. *European Journal of Operational Research* 97:409–38.

Daganzo, C. F. 1999. *Logistics Systems Analysis*. Berlin: Springer.

Daskin, M. S. 1985. Logistics: An overview of the state of the art and perspectives on future research. *Transportation Research B* 19B:383–98.

Dökmeci, V. F. 1973. An optimization model for a hierarchical spatial system. *Journal of Regional Science* 13:439–51.

Eilon, S., Watson-Gandy, C. D. T., and Christfides, N. 1971. *Distribution management: Mathematical modelling and practical analysis*. London: Griffin.

Hagget, P. 1967. An extension of the Horton combinatorial model to regional highway networks. *Journal of Regional Science* 7:281–90.

Higginson, J. K. 1993. Modeling shipper costs in physical distribution analysis. *Transportation Research A* 27A:113–24.

Horton, R. E. 1945. Erosional development of streams and their drainage basins; hydrophysical approach to quantitative morphology. *Bulletin of the Geological Society of America* 56:275–370.

Ieda, H. 1997. Modeling and solving of hierarchical transport systems such as hub-spokes/point-to-point and/or consolidated/direct delivery in the uniform boundless plane. *Infrastructure Planning Review* 14:773–82 (in Japanese).

Ieda, H., Sano, K., and Tsuneyama, S. 1992. A macroscopic collection/delivery model of goods transport and its application on an inter-carrier co-operative cargo logistics. *Infrastructure Planning Review* 10:247–57 (in Japanese).

Ishiguro, K., Sakurada, T., and Inamura, H. 2000. A location model of interregional freight complexes applying cost minimization principle with the scale economy. *Infrastructure Planning Review* 17:693–700 (in Japanese).

Kasahara, K., and Furuyama, M. 1998. A study of the length of the minimum spanning tree and the hierarchical tree. *Journal of Architecture, Planning and Environmental Engineering* 504:155–61 (in Japanese).

Kijmanawat, K., and Ieda, H. 2004. Multilevel hierarchical network design: Formulation and development of M-GATS algorithm. *Journal of Infrastructure Planning and Management* 751:139–50.

Klose, A., and Drexl, A. 2005. Facility location models for distribution system design. *European Journal of Operational Research* 162:4–29.

Kurita, O. 1999. A mathematical model on the optimal number of urban facilities: Empirical analysis on the number of the wards of the ordinance-designated cities in Japan. *Journal of Architecture, Planning and Environmental Engineering* 524:169–76 (in Japanese).

Langevin, A., Mbaraga, P., and Campbell, J. F. 1996. Continuous approximation models in freight distribution: An overview. *Transportation Research A* 30A:163–88.

Lösch, A. 1940. *Die räumliche Ordnung der Wirtschaft*, Jena, Germany: Fischer.

Matsumoto, S. 1990. Trade-off between logistic cost and customer service for goods distribution in an urban area. *Journal of the Japan Society of Civil Engineers* 413:31–38 (in Japanese).

Narula, S. C. 1984. Hierarchical location-allocation problems: A classification scheme. *European Journal of Operational Research* 15:93–99.

Ogura, M. 1999. *Ogura Masao Keieigaku*. Nikkei BP (in Japanese), Tokyo.

Okabe, A., and Okunuki, K. 1993. Analysis and optimization of spatial hierarchy. *Proceedings of the Fifth RAMP Symposium*, 127–40 (in Japanese).

Okunuki, K., and Okabe, A. 1995. Optimization of successively inclusive hierarchical facilities on a plane. *Journal of the City Planning Institute of Japan* 30:565–70 (in Japanese).

Sadahiro, Y., and Okabe, A. 1994. Method for describing hierarchical structure of urban facilities. *Geographical Review of Japan* 67A:225–35 (in Japanese).

Shreve, R. L. 1996. Statistical law of stream numbers. *Journal of Geology* 75:178–86.

Stephan, G. E. 1988. The distribution of service establishments. *Journal of Regional Science* 28:29–40.

Suzuki, T. 1990. A note on the optimal hierarchy system of the facilities. *Journal of the City Planning Institute of Japan* 25:331–36 (in Japanese).

Suzuki, T. 2000. A test of the size-density hypothesis in *p*-median problems. *Journal of Architecture, Planning and Environmental Engineering* 532:171–76 (in Japanese).

Suzuki, T., and Kawaguchi, A. 1998. Economy of scale in transportation and optimal partitioning in a one-dimensional city. In *Abstracts of the 1998 Fall National Conference of Operations Research Society of Japan*, 24–25 (in Japanese).

Tokunaga, E. 1996. The composition of drainage network in the Toyohira-River basin and valuation of Horton's first law. *Geophysical Bulletin of the Hokkaido University* 15:1–19 (in Japanese).

Tokunaga, Y., Okada, R., and Suda, H. 1995. The model of distribution and transportation route for express delivery service. *Infrastructure Planning Review* 12:519–24 (in Japanese).

Watanabe, D. 2006. Mathematical study on the network pattern analysis using proximity relations and the optimization of transportation system. Doctoral dissertation thesis, Graduate School of Systems and Information Engineering, University of Tsukuba (in Japanese).

Watanabe, D., and Suzuki, T. 2000. A study on the hierarchical collection/distribution system with economies of scale. *Journal of the City Planning Institute of Japan* 35:1027–32 (in Japanese).

4 A Study of the Route-Memorizing Mechanism

*Experiments through Computer-Aided Walking Simulation**

Teruhisa Kamachi,† Yasushi Asami,‡ and Atsuyuki Okabe§

4.1 INTRODUCTION

We often consult maps when we visit a city for the first time. It is frequently the case that we do not look at the map continuously, but repeatedly consult it to memorize the way to a certain point and then walk to that point. By analyzing the memorizing process for a route, we can see how to construct easy-to-comprehend road networks and where to locate road maps on the streets. Another important question is the memorizing unit for routes. Can we quantitatively analyze memory loads for route segments? To answer these questions, experiments through computer-aided walking simulation (Kamachi, 1988) have been conducted and the results analyzed.¶

A variety of studies related to walking routes have been performed in the field of spatial cognition. (For the review of studies in spatial cognition on a large spatial scale, see Siegel and White (1975).) For example, Thorndyke and Hayes-Roth (1982) analyzed knowledge acquisition from maps, and showed that maps are an effective device for revealing the relative spatial relationship. Cousins et al. (1983) analyzed children's ability of route finding, and Passini (1980) summarized concepts for route-finding

---

* This is a translation of a paper by Kamachi, T., Asami, Y., and Okabe, A., "A Study on the Route-Memorizing Mechanism: Experiments through Computer-Aided Walking Simulation," *Papers on City Planning* 23 (1988): 7–12 (with minor modifications to improve presentation).

† C&C Data Business Office, NEC Corp. (in 1988). Currently affiliated with So-net Entertainment Corp.

‡ Department of Urban Engineering, the University of Tokyo (in 1995). Currently affiliated with the Center for Spatial Information Science, the University of Tokyo.

§ Department of Urban Engineering, the University of Tokyo (in 1995 and now in 2008).

¶ PC9801VX was used for the experiment. Kamachi coded with C language. For details, see Kamachi (1988).

mechanisms. Moar and Carleton (1982) also tried to analyze route-memorizing mechanisms. The existing studies largely use the actual space for their experiments. A variety of factors therefore interfere with the results, and the pure effect of those factors cannot be precisely analyzed. Even in an apparently featureless field, it was found that subjects utilized subtle differences in the actual space as landmarks (Heft, 1979). To overcome this difficulty, we experiment with completely featureless streets without landmarks on a computer. The final goal of our study is to quantify how much people can memorize each element in memorizing a route. Experiments are based on the preliminary study by Asami et al. (1988), and the results are analyzed subsequently.

A brief overview of the experiment is described in Section 4.2, and preliminary studies based on postexperiment interviews with subjects are presented in Section 4.3. Regression analyses for time required for walking through, time required for reading maps, etc., are presented in Section 4.4. It is shown that the frequency of consulting maps differs depending on the pattern of the road. Section 4.5 is devoted to the analysis of memory loads for route segments based on the discussion in Section 4.3. The results show that the memory loads are identified as route segments classified by the pattern of intersection shape, that on average maps are consulted every two turnings, and that the memory capacity ranges from half the average to three times the average. Section 4.6 concludes the chapter with discussion and suggestions for future research.

4.2 OVERVIEW OF THE EXPERIMENT

Experiments were conducted in July 1987 in the research unit room in the Department of Urban Engineering at the University of Tokyo. Subjects were between 20 and 35 years old (the average being 22), and were 24 male students or graduates of the Faculty of Engineering at the University of Tokyo who did not major in psychology. After the experiment had been described to them, subjects were asked to walk through a simple road network (shown in Figure 4.1) in a walking simulator on a PC9801VX machine to practice the key operation. First, the map showing the present location, the facing direction, and the destination appeared on a CRT display. By pushing the key labeled "quit map," subjects obtained a 3D perspective (as in Figure 4.2) of the road from the present location. The roads were indicated by walls on both sides. By pushing the arrow keys, the subjects could modify the picture a little so that they felt as though they were walking on the road. By pushing the arrow keys many times, the subjects could gradually walk through to reach the destination. A blue pole was standing at the destination point, which could be seen from the place where the destination was visible, which helped subjects to know the location of the destination on a CRT display. Subjects were allowed to consult the map at any time during the experiment by pushing a key labeled "map," which showed the present location, the facing direction, and the destination.

Subjects were instructed to reach the destination as soon as possible in the experiment. Subjects walked four road patterns (shown in Figure 4.3) in addition to dummy patterns.* Four routes were the same in terms of distance, places to turn, and the

* Dummy patterns were a curvy route with different locations to turn. This was inserted between the road patterns for analysis to avoid the subject spotting that the correct routes in the four patterns were the same shape.

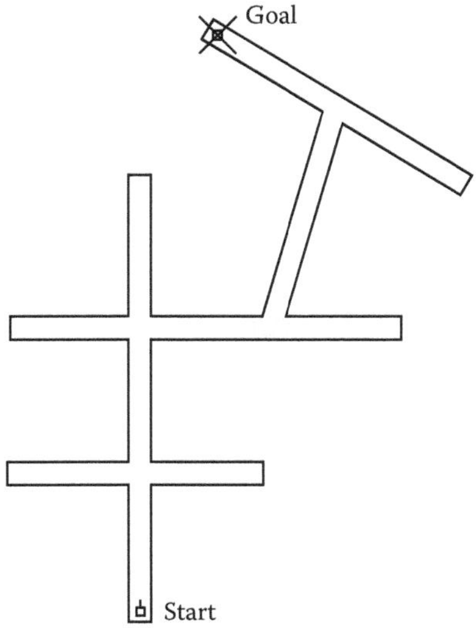

FIGURE 4.1 Map view (test pattern).

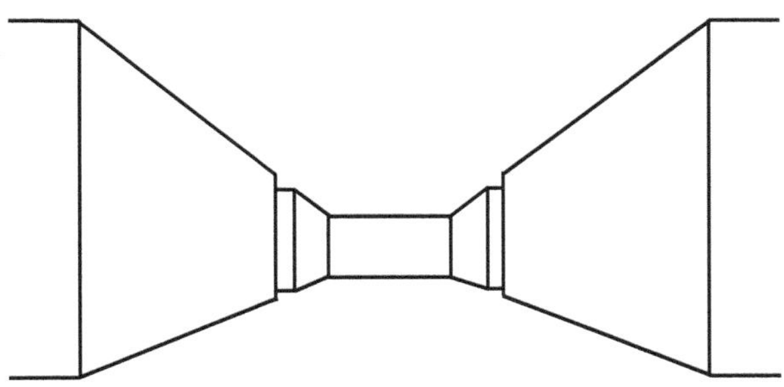

FIGURE 4.2 Road view.

direction they had to turn in to reach the destination. The aim of this design was not explained to the subjects, but the interviews after the experiment revealed that six subjects (out of 24 persons) had already discovered this fact through the experiment. For each pattern, there was only one correct route (shortest path), and only one kind of intersection appeared. Pattern I in Figure 4.3 consisted of only L-shaped intersections, and if the subjects found out this fact, then they did not need to consult maps to reach the destination. Pattern II consisted of only T-shaped intersections, and which direction to turn in was the only information they needed to reach the destination. Pattern III consisted of only T-shaped intersections to turn in one direction, and the number of intersections to the next turning point was the only information they needed to reach the destination. Pattern IV consisted of cross-intersections, and both the direction to turn in and the number of intersections to the next turning point had to be known in order to reach the destination. All the key operations of the subjects were recorded with time information.

After the experiment, subjects were asked questions about their strategy for memorizing the route in a 10-minute interview. Since the descriptions given by the

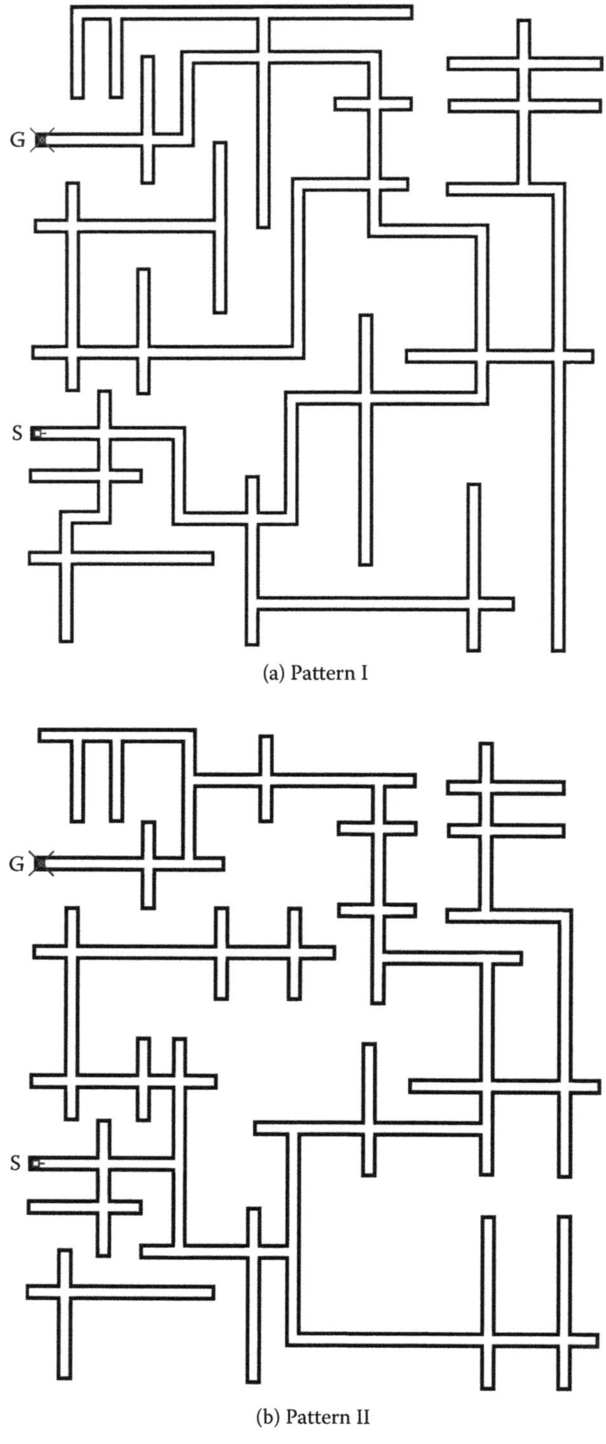

FIGURE 4.3 Four patterns used in the experiment.

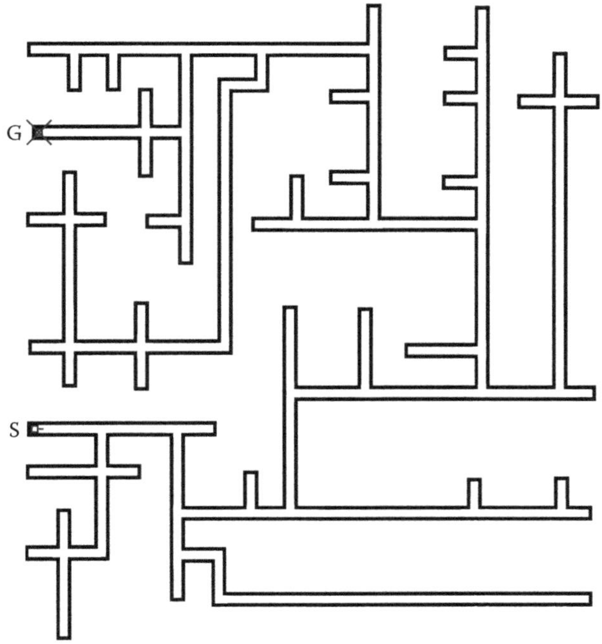

(c) Pattern III

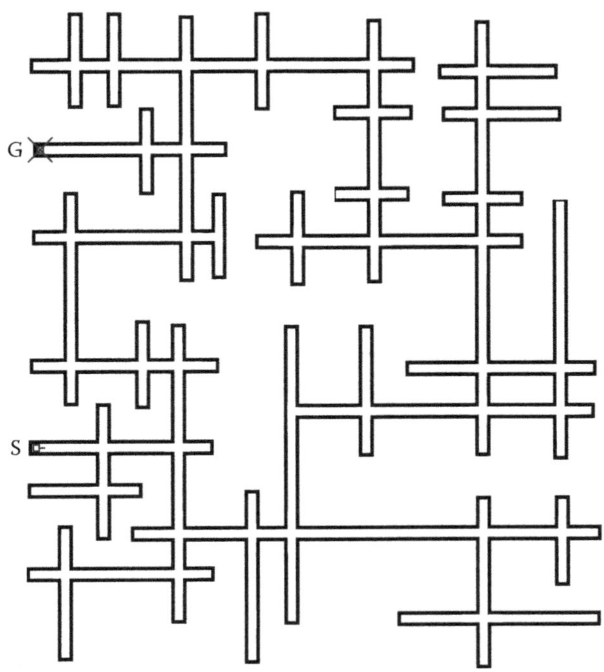

(d) Pattern IV

FIGURE 4.3 *(Continued)*

subjects were consistent with the results of key operations, the data were analyzed according to the subjects' statements.

4.3 PRELIMINARY STUDIES CONCERNING ROUTE MEMORIZATION: INTERVIEWS WITH SUBJECTS

Subjects do not always memorize the entire route by consulting a map. Rather, subjects tend to memorize part of the route and walk it before looking at the map again (Thorndyke and Stasz, 1980). Based on the interviews with subjects, strategies for memorizing the route were identified for each subject. There are seven strategies, as follows, with the list of patterns applicable in parentheses:

1. Memorizing the pair consisting of the direction to turn and the number of intersections to the next turning point (I, II, III, IV)
2. Memorizing the direction to the turn (I, II)
3. Memorizing the number of intersections to the next turning point (I, III)
4. Going along the road by finding out that all the intersections to turn into were L-shaped intersections (I)
5. Memorizing the number of intersections to the next turning point and the general shape of the route (I, II, III, IV)
6. Memorizing the general shape of the route (I, II)
7. Memorizing the codes for all the intersections (including the intersections to pass over), such as straight, right, or left (I, II, III, IV)

Most of the subjects utilized strategy 1 or its modifications, 2, 3, and 4. We can therefore assume that the path segment to the next intersection to turn into can be regarded as a unit for memorizing routes.

4.4 REGRESSION ANALYSES

Regression analyses were conducted for net time by seconds required for walking through to the destination (TT: time from the starting point to the goal minus time required for machine computation, such as controlling the screen), total time by seconds required for consulting map (MT), and the number of times to consult the map (KS) as dependent variables. Independent variables are dummy variables for map pattern (M2: dummy variable for pattern II; M3: dummy variable for pattern III; M4: dummy variable for pattern IV), dummy variables for order of appearance (Z1: dummy variable for second time; Z2: dummy variable for fourth time; Z3: dummy variable for sixth time), the number of times the subject got lost on the way (NL: the number of times that the subject deviated from the correct route), and the variable that the subject found out that the correct routes were the same shape (SM = x: when the subject reported that he had found out, $x = 0.25$ for fourth map, $x = 0.50$ for sixth map and $x = 1.0$ for eighth map).*

* If the subject finds out that the correct route is the same in shape, then he can easily walk through by using the last memory. To control this effect, the variable was weighted as described.

TABLE 4.1
Results of Regression Analyses

$$TT = 77.17 + 19.38{*}M2 + 29.99{*}M3 + 31.39{*}M4 + 18.71{*}Z1 + 0.98{*}Z2 + 4.90{*}Z3 + 26.43{*}NL - 2.41{*}SM$$

$$(7.95)\quad (1.88)\qquad (2.87)\qquad (3.02)\qquad (1.75)\qquad (0.09)\qquad (0.48)\qquad (4.98)\qquad (-0.17)$$

$$R = 0.6321$$

$$MT = 22.68 + 10.47{*}M2 + 12.76{*}M3 + 16.69{*}M4 + 1.98{*}Z1 + 2.12{*}Z2 + 0.13{*}Z3 + 9.37{*}NL - 2.00{*}SM$$

$$(5.45)\quad (2.37)\qquad (2.85)\qquad (3.74)\qquad (0.43)\qquad (-0.47)\qquad (0.03)\qquad (4.13)\qquad (-0.32)$$

$$R = 0.5986$$

$$KS = 3.214 + 0.585{*}M2 + 2.420{*}M3 + 1.933{*}M4 + 0.290{*}Z1 + 0.082{*}Z2 - 0.166{*}Z3 + 1.328{*}NL - 0.450{*}SM$$

$$(4.77)\quad (0.82)\qquad (3.34)\qquad (2.68)\qquad (0.39)\qquad (0.11)\qquad (-0.23)\qquad (3.61)\qquad (-0.45)$$

$$R = 0.5585$$

Note: *t* values in parentheses. Number of cases is 96. R is the correlation coefficient.

The results of the analyses are shown in Table 4.1. The results show that:

1. Regression analysis for net time required, TT, identified three significant variables (M3, M4, NL). It was found that pattern III or pattern IV required 30 s more than pattern I, and an additional 25 s when the subject got lost on the way.
2. Regression analysis for map consultation time, MT, identified four significant variables (M2, M3, M4, NL). It was found that an additional 10 s, 13 s, and 17 s were required for pattern II, pattern III, and pattern IV, respectively, compared with pattern I for map consultation, and an additional 9 s for map consultation when the subject got lost on the way.
3. Regression analysis for the number of map consultations, KS, identified three significant variables (M3, M4, NL). It was found that subjects tended to consult maps for patterns III and IV twice as much as for pattern I, and one additional map consultation was required when the subject got lost on the way.

These results show that subjects consulted the map for the patterns requiring the number of intersections to the next intersection to turn. In other words, memorizing the number placed a higher memory load on the subjects. The constant term for the regression equation for KS reveals that when it is necessary to remember a number, subjects can memorize up to two turns, while when it is not necessary to memorize the number, subjects can memorize up to three turns on average.

4.5 QUANTIFICATION OF ROUTE MEMORY LOADS

Using only the subjects taking the major strategies, 1 to 3, described in Section 4.3 for memorizing the route, we calculated the route memory loads. We also confined subjects to those who walked through the correct route in order to discard the data for subjects' possible vague memory at the parts where they went astray. The number of effective subjects was nineteen. Since it was known that the route segment from a point after turning at an intersection to the next turn at an intersection was the unit to memorize, the route segments were classified in the following classes. The numbers in parentheses are the element numbers to be used below.*

(1) D1: Turning at the first intersection
(2) D2: Turning at the second intersection
(3) D3: Turning at the third intersection
(4) G1: Passing through the first intersection
(5) G2: Passing through the second intersection
(6) D: Turning at the end T-shaped intersection
(7) N1: Turning at the first intersection along the road
(8) N2: Turning at the second intersection along the road

These classifications were made in the light of the subjects' strategies. For example, if the subject described his strategy as memorizing both the direction to turn and the number of intersections to the next intersection, then even if the actual turning intersection was a T intersection and did not need the number, D1 to D3 were assigned instead of D. G1 or G2 was assigned when the subject consulted the map more than once on the same straight road segment before the intersection to turn. N1 or N2 was assigned when the subjects reported that they did not need the directional information for pattern III. How can we quantify the memory loads? To the authors' knowledge, no paper exists that deals with this issue (except for that by Asami et al. (1988)). The following *method of minimizing memory capacity variation* is formulated and the memory loads are estimated.

Let b_j be the memory load for element j. $b = (b_j: j = 1, \ldots, 8)$ is the vector of memory loads. Suppose that from the time a subject consulted a map to the time he next consulted the map, he walked $x(j)$ units for element j. Then assuming the additivity, the sum, $\Sigma_j b_j x(j)$ can be regarded as the subject's memory capacity for this path. We may find out average memory loads using this kind of value for all the paths. The first consultation of a map may differ from other consultations, for it is the first time of seeing the map, and therefore the subjects may grasp the entire route in addition to memorizing the immediate walking part of the route. The last consultation of the map may also differ from other consultations, for it may not reach the maximum capacity of the subject's memory. For this reason, the first and last paths are ignored in the following analysis. Suppose that there are n usable paths for all subjects. Let $X = (x_{ij}: i = 1, \ldots, n, j = 1, \ldots, 8)$. Denoting the average memory capacity of subjects by c, then the total memory loads for any path can be expected to be close to c, implying that Xb is close to $c\mathbf{1}$, where $\mathbf{1}$ is the vector whose elements are all 1s. Depending on the individual difference in memory and the situation of paths, this may fluctuate. Thus, it seems appropriate to estimate

* Only one N3 (turning at the third intersection along the road) was found in all 193 path data (path consisted of a point from map consultation to the next map consultation), but the sample of one was too small to get reliable results, and hence it was omitted in the subsequent analysis.

TABLE 4.2
Memory Load for Each Element (ignoring individual difference in memory capacity)

Element	Memory Load	Element	Memory Load
D1	0.1039	G2	0.1654
D2	0.1394	D	0.0840
D3	0.1233	N1	0.0654
G1	0.1122	N2	0.2066

Note: To make Table 4.2 comparable with Table 4.3, the memory load values are standardized such that the sum is equal to 1.

the memory load vector, b, by minimizing the variation, $(\mathrm{X}b - c\mathbf{1})^{\mathrm{T}}(\mathrm{X}b - c\mathbf{1})$, where superscript T is the transposition operator. If we define relative memory load vector, a, by b divided by c, i.e., $a = b/c$, then a can be derived by solving the following problem:

$$\min_a (\mathrm{X}a - \mathbf{1})^{\mathrm{T}}(\mathrm{X}a - \mathbf{1})$$

The first-order condition yields the following solution for a:

$$a = (\mathrm{X}^{\mathrm{T}}\mathrm{X})^{-1}\mathrm{X}^{\mathrm{T}}\mathbf{1}$$

The memory load derived with this procedure is shown in Table 4.2.

1. Memory load for turning at the first intersection (D1 and N1) is smaller than that for turning at the second intersection or more (D2, D3, N2).
2. The element with the smallest memory load is N1, with memory load 0.0654. This element represents the case in which a subject turns at the first intersection without the need to memorize the turning direction. In other words, the memory load for this element is almost equivalent to memorizing 1. The element with the second smallest memory load is D, which is the case where a subject should turn at the end T intersection. In this case, the subject does not have to memorize the number of intersections; all he needs is the direction to turn. In other words, the memory load for this element is almost equivalent to memorizing "right" or "left." In information theory, information quantity for both elements is 1 bit (i.e., information of 0 or 1). Accordingly, the similar memory load values for both cases can be regarded as a natural result.
3. The order of the magnitude of memory load for G1 and G2 appears appropriate. For G1, a subject has to remember passing at the first intersection, but for G2, the subject has to remember passing two intersections. Accordingly, the subject has to remember more information for G2.

The results in Table 4.2 may not be complete. To improve the analysis, a more detailed analysis is conducted taking into consideration the subjects' variation in memory capacity. The memory capacity of subjects and the memory load for each element are derived as follows. Let K be the number of subjects. Let c_k and n_k be subject k's memory capacity and number of paths, respectively ($k = 1, \ldots, K$). Then Xb is expected to be close to $C = (c_1\mathbf{1}(n_1)^{\mathrm{T}}, c_2\mathbf{1}(n_2)^{\mathrm{T}}, \ldots, c_K\mathbf{1}(n_K)^{\mathrm{T}})^{\mathrm{T}}$ by argument similar to the above, where $\mathbf{1}(m)$ is m-dimensional vector with all elements equal to 1. To standardize the memory load, we assume

$$\mathbf{1}^{\mathrm{T}}b = 1$$

The problem to solve is therefore formulated as

$$\min_{b,c} (\mathrm{X}b - C)^{\mathrm{T}}(\mathrm{X}b - C) \quad \text{s.t. } \mathbf{1}^{\mathrm{T}}b = 1$$

where $c = (c_k\!: k = 1, \ldots, K)$.

Table 4.3 shows the resulting values of memory load, and Table 4.4 shows the estimated memory capacity of subjects.

1. The result in Table 4.3 appears to be more reliable than that in Table 4.2. In particular, order in magnitude of memory loads for S1, D2, and D3 is now acceptable. It shows that when a subject has to memorize the direction and the number of intersections to the next intersection to turn, the more the intersections, the harder it becomes to memorize (the more the memory load becomes). The difference between D1 and D2 is larger than that between D2 and D3 in terms of memory load, which implies that the memory load of turning the first intersection and that of turning the second or later intersection is very different. When we need to turn at the first intersection, we just memorize that we have to turn at the *next* intersection, but when we need to turn at the second (or third) intersection, we may not memorize "turn at the *next to next* (*to next*) intersection." This suggests that the ways of memorizing them may be fundamentally different. The result of the experiment seems to support this idea.
2. As expected, element D, which should need only the direction to turn, exhibits the smallest memory load.

TABLE 4.3
Memory Load for Each Element
(taking into account individual difference in memory capacity)

Element	Memory Load	Element	Memory Load
D1	0.0743	G2	0.0822
D2	0.1052	D	0.0679
D3	0.1092	N1	0.2201
G1	0.0978	N2	0.2432

TABLE 4.4
Memory Capacity of Subjects (taking into account individual difference in memory capacity; the unit is the same as in Table 4.3)

Subject ID	Memory Capacity	Subject ID	Memory Capacity	Subject ID	Memory Capacity
2	0.1836	12	0.2662	18	0.2027
3	0.1379	13	0.2342	19	0.1654
4	0.1221	14	0.1375	20	0.3110
8	0.1153	15	0.2043	22	0.6519
9	0.4022	16	0.1879	23	0.2787
10	0.1373	17	0.1313	24	0.1648
11	0.2476				

Note: Average = 0.2254; standard deviation = 0.1273.

3. Table 4.4 reports the memory capacity of subjects. The average memory capacity is 0.2254. Accordingly, by consulting a map once, subjects can walk through three D1 or D elements, two D2, D3, G1, or G2 elements, and one N2 element on average. In other words, the average subject can remember two pairs of numbers and directions through one map consultation. Minimum memory load was found to be 0.1153 for subject 8, and maximum memory load was found to be 0.6519 for subject 22. A person with the lowest memory capacity has about half the capacity of the average person, and needs to consult the map every time he turns at an intersection. On the other hand, a person with the highest memory capacity has about three times the capacity of the average person, and can memorize a path consisting of six turns. We often encounter arrow marks for site guidance (such as in exhibition halls and funeral halls). In such cases, almost all the intersections to turn are indicated by arrow marks. The result of the experiment shows that this produces the situation where even the person with the lowest memory capacity does not go astray.
4. The memory loads for N1 and N2 are unexpectedly large. The order of their magnitude seems reasonable, but the values are somewhat larger than the memory loads of D1, D2, and D3. In the case of N1 and N2, the subjects need not memorize the direction to turn. It therefore seems reasonable to have the memory load for N1 smaller than that for D1, and that for N2 smaller than that for D2, but the result is contrary to this expectation. This may be because of the small number of cases for these elements (13 for N1 and 4 for N2). In the interviews after the experiment, some subjects said that the pattern consisting of elements N1 to N3 (pattern III) was more difficult than the pattern consisting of elements D1 to D3 (pattern IV). This may be because pattern IV is a much more familiar pattern. It may also be that memorizing pairs of numbers and directions may be easier to do than a series of numbers alone, even though the

amount of information is greater. If subjects memorize pairs of numbers and directions, then they can easily recognize an error when they go the wrong way. On the other hand, if the subjects memorize only numbers, then they may face difficulty in recognizing an error when they go the wrong way. Because of this low possibility of recognition, subjects may choose to consult the map more frequently. If this proposition is correct, then it may be a good strategy to memorize pairs of numbers and directions. Why the memory load for N1 and N2 is so large is a question that should be examined in the future. Incidentally, this result conforms to the result of regression analysis in that the pattern requiring numbers only requires more time and more map consultations.

5. The memory load for "one more" (G1) appears to be smaller than that for "two or more" (G2), but the results show the opposite tendency. The number of observations for G2 (which was 6) is much smaller than that for G1 (which was 26), and therefore the conclusion is not decisive enough. It can be said that the memory loads for G1 and G2 are not so different. G1 or G2 (i.e., consulting a map before turning) will be taken when subjects are not confident about which intersection to turn at. Accordingly, similarity in values of memory loads is a natural result.

4.6 CONCLUSION

This study conducted a walking experiment with a walking simulator on computer, and the results are analyzed quantitatively. Results show that: (1) Patterns III and IV (which required subjects to memorize the number of intersections to the turning point) require more time and more map consultations than other patterns. (2) The memory load for the element requiring only the direction to turn (D) is the smallest. (3) The memory load for turning at the first intersection and that for turning at the second or later intersections are quite different. (4) On average, information obtained from consulting a map once can cover two turns. (5) When subjects do not have to memorize the direction to turn, the number of map consultations tends to be larger. (6) The individual difference in memory capacity is significant. The minimum capacity was about half of the average, and the maximum capacity was about three times the average.

Several issues should be studied further in the future. The constant memory capacity hypothesis, on which all these analyses are based, should be examined further. Similar studies should be undertaken by the collection of more samples to improve the reliability of the analysis. Other factors, such as landmarks, which influence memory of a route, must be examined. Studies along these lines will be published in other chapters in the future.

ACKNOWLEDGMENT

The authors are indebted to anonymous reviewers for valuable comments, which are gratefully acknowledged.

REFERENCES

Asami, Y., Kamachi, T., Machida, N., Nishie, F., Okabe, A., Sasaki, M., Tanaka, T., Teraki, A., and Yamada, T. 1988. *A preliminary analysis of the route memorization mechanism.* Discussion Paper 34, Department of Urban Engineering, the University of Tokyo.

Cousins, J. H., Siegel, A. W., and Maxwell, S. E. 1983. Way finding and cognitive mapping in large-scale environments: A test of a development model. *Journal of Experimental Child Psychology* 35:1–20.

Heft, H. 1979. The role of environmental features in route-learning: Two exploratory studies of way-finding. *Environmental Psychology and Nonverbal Behavior* 3:172–85.

Kamachi, T. 1988. Environmental psychological analysis of memorizing walking route. Master's thesis, Department of Urban Engineering, the University of Tokyo.

Moar, I., and Carleton, L. R. 1982. Memory for routes. *Quarterly Journal of Experimental Psychology* 34A:381–94.

Passini, R. 1980. Wayfinding: A conceptual framework. *Man-Environment Systems* 10:22–30.

Siegel, A. W., and White, S. H. 1975. The development of spatial representations of large-scale environments. In Reece, H. W. (ed.), *Advances in child development and behavior.* New York: Academic Press, pp. 9–55.

Thorndyke, P. W., and Hayes-Roth, B. 1982. Differences in spatial knowledge acquired from maps and navigation. *Cognitive Psychology* 14:560–89.

Thorndyke, P. W., and Stasz, C. 1980. Individual differences in procedures for knowledge acquisition from maps. *Cognitive Psychology* 12:137–75.

5 Artificial Neural Network Model Estimating Land Use Change in the Southwestern Part of Nagareyama City, Chiba Prefecture[1]

Fumiko Ito and Akiko Murata

5.1 INTRODUCTION

Land use change is one of the essential fields in urban analysis. The question is, what changes are caused by what factors and in what structure? This question has been the theme of a number of studies, which are broadly grouped into two: those focusing on the relationship between various forms of land use and their causal factors, and those attempting to develop estimation models for such changes. Many studies in the former group ascribe the changes to a major factor such as land use zoning (Yoshikawa et al., 1990; Chishaki et al., 1990). Not a few studies in the latter group, land use changes, are estimated on the basis of one factor—less frequent cases of adjacent land use. Since the 1990s, new models have been attempted by what is known as the Markov chain model (Osaragi and Masuda, 1995), the genetic algorithm (Takizawa et al., 1997), and the cellular automaton model (Arai et al., 1999), all with the background of development and diffusion of electronic data in recent years.

Factors for changing land use have been categorized into three types (Kawakami and Honda, 1994): socioeconomic factors, natural factors, and regulation factors of urban planning. The aforementioned studies discuss various other factors. The past models were required to explain the land use change with fewer factors. The actual picture of land use, however, shows that various factors are closely intertwined, and thus it is advisable to compose effective models that consist of these various factors.[2] Not many models of land use change have, however, reflected such varying factors properly until now. To include many factors, models should have flexible structure. Fortunately, the neural network (NN) model has the flexibility to determine its model structure through its own learning process. Edamura and Kawai (1992) applied NN to land use change, and presented the percentage of total floor area by blocks in residential, commercial, and industrial quarters. Our study also uses the detailed digital information data to estimate land use changes for each 10 m grid in an attempt to construct a more detailed model.

In the past, the NN model was regarded as highly adaptable but difficult in terms of revealing its inner structure. As it is important to identify the structure of land use change, or the degree of the effect of each factor on the change, our study also aims to clarify the inner structure of the model and thus estimate the degree of effect of each factor.

This study aims to estimate quantitatively the effect of various factors by taking into full account the factors for land use change and constructing a detailed on-the-spot model of NN to investigate the structure of the changes.

This chapter has the following sections: Section 5.2 presents the areas under study and the data used. Section 5.3 quantitatively grasps the changes in land use over a decade (1984–1994) in the study area. Section 5.4 extracts factors that cause changes in land use and develops an NN model that would output land use change after 10 years. Section 5.5 estimates the degree of the effect of each factor that causes changes in land use by clarifying the inside of the model developed in Section 5.4. Section 5.6 summarizes the whole process.

5.2 STUDY AREA AND DATA USED

5.2.1 Study Area

The area for this study is the southwestern part of Nagareyama City in Chiba Prefecture (Figure 5.1). Thirty kilometers from the center of Tokyo, this area has witnessed the construction of new stations with the opening of the Tsukuba Express line in 2005 (linking Tsukuba Academic New Town with Tokyo) and the launching of a land readjustment project.

5.2.2 Data Used

The data used in this study include 1984/1994 "Detailed Digital Information" (Geographical Survey Institute), 10 m grid land use data (Table 5.1: Land Use Classification of the Then Ministry of Construction), altitude and gradient by means of the 100 m grid and data on the enforcement area for the land readjustment project, station points data/road line data in "Digital Map 2500" (Geographical Survey Institute), "Nagareyama City Statistics" for 1984 and 1994, population data for town of Nagareyama City, and Nagareyama City planning map 1983–1992.

5.3 QUANTITATIVE ANALYSIS OF LAND USE CHANGE

5.3.1 Changes in Land Use

To gain an overview of the land use change, we compared the situation of land use in 1984 with that of 1994 in the area under study. This area has 14 types of land

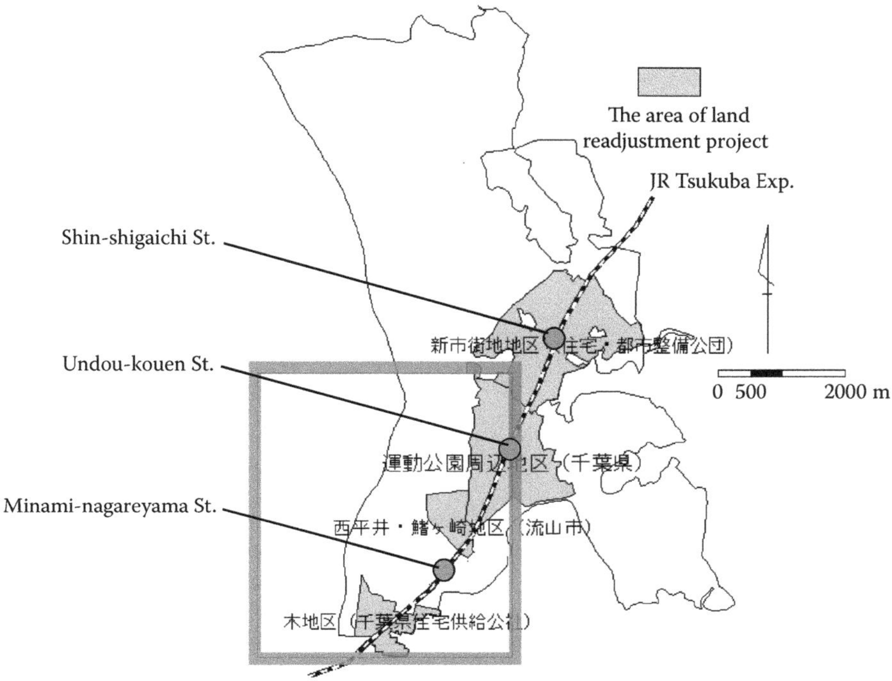

FIGURE 5.1 Study area in Nagareyama City.

use, out of 16 classes of land use drawn up by the former ministry of construction. Except for four less-changed types, such as "river, lake, and marsh,"[3] the remaining ten types are further classified into six land use categories: forest land, farmland, vacant, industrial, residential, and commercial (Table 5.1). Figure 5.2 shows the land use in the study area classified into six categories for every 10 m grid. The figure shows a steady conversion from vacant land into residential land in the southern part of the area, and from forest land and farmland into residential and vacant land in the northeastern part.

These changes can be quantitatively confirmed in Table 5.2, in terms of the number of grids over the changed or unchanged land use for the years 1984 to 1994, and their corresponding percentages (with 1984 as 100). The table shows that the relatively higher percentages of unchanged (retained) land use are 96% for residential, 88% for commercial, and 84% for farmland. In contrast, the relatively higher percentages of changed land use are 51% for vacant, 32% for industrial, and 27% for forest land. Relatively high percentages of conversion are 33% from vacant into residential, 28% from industrial into commercial land, and 17% and 8% from forest land into vacant and residential land, respectively.

5.3.2 Analysis

These facts show that conversion over the 10 years was heavier from forest land into vacant land, from farmland into vacant and residential land, from vacant land

TABLE 5.1
Land Use Classifications by MLIT and Categories Used in This Study

Category of Land Use	Classifications of Land Use
Forest land	1. Forest, wasteland
Farmland	2. Rice fields
	3. Fields and other farmland
Vacant	4. Reclamations land
	5. Vacant
Industrial	6. Industrial land
Residential	7. Low-rise housing
	8. Densely built low-rise housing
	9. High-rise housing
Commercial	10. Commercial and business district
Exclude from analysis	11. Roads
	12. Parks and green space
	13. Land for public facilities
	14. River, lake, and marsh
Does not exist in study area	15. Others (base, imperial estate)
	16. Sea

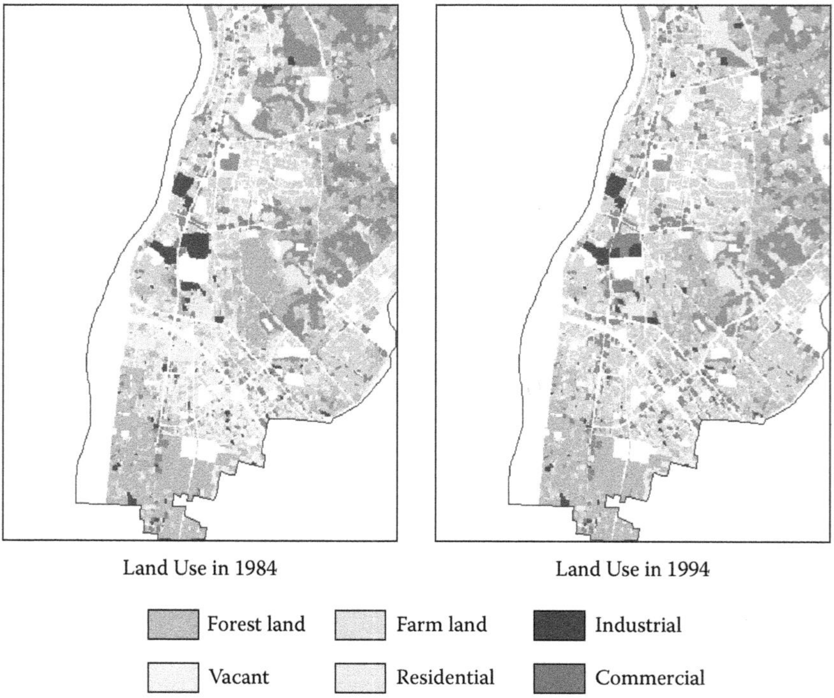

FIGURE 5.2 Land use.

TABLE 5.2
Changes in Land Use between 1984 and 1994

Land Use of 1994	Land Use of 1984: Forest land	Farmland	Vacant	Industrial	Residential	Commercial	Total
Forest land	6,082		45				6,127
	73%		0%				
Farmland	59	15,849	792	11	45	13	16,769
	1%	84%	7%	1%	0%	0%	
Vacant	1,380	1,807	5,370	38	600	221	9,416
	17%	10%	49%	2%	4%	7%	
Industrial	14	52	75	1,095	6	41	1,283
	0%	0%	1%	68%	0%	1%	
Residential	674	904	3,647	18	16,073	118	21,434
	8%	5%	33%	1%	96%	4%	
Commercial	72	288	988	441	51	2,808	4,648
	1%	2%	9%	28%	0%	88%	
Total	8,281	18,900	10,917	1,603	16,775	3,201	59,677

Upper: Number of grids
Lower: %

into residential and commercial land, and from industrial into commercial land. An overall picture is a strong trend of conversion into residential land. While generally seen in those years in many other suburban cities like the area under study, this trend vindicates our analysis of the study area. The next section will extract factors and develop a land use change model consisting of those factors to clarify the structure of the change.

5.4 LAND USE CHANGE MODEL

5.4.1 Extraction and Selection of Factors Causing Changes in Land Use

Ten factors causing conversions in land use were extracted from the prior studies quoted in Section 5.1. The factors were categorized by condition into socioeconomic conditions (present state of land use, adjacent land use, changes in population); traffic conditions (distance from stations, time from the center of Tokyo, distance from trunk roads); urban planning conditions (land use zoning, floor-area ratio); and natural conditions (altitude, gradient). "Number of years after the readjustment project completed" has been added as another factor, since its effect is relevant to the model's estimation of future land use in this area, where a large-scale land readjustment project is scheduled. It comes to 11 factors in total.

For each of these eleven factors, the grid data have been computed from the data sources listed in Section 5.2.2 as follows. The area under study has been divided by geographic information systems (GIS) software into 100 m grids to obtain attribute values of nine features: distance from stations, time from the center of Tokyo, distance from trunk roads, changes in population, altitude, gradient, designated land use, designated capacity ratio, and years after the land readjustment project completed. The NN models for land use in this study are intended to estimate the situation of land use 10 years ahead on the basis of the present land use as well as other factor values. The present land use of 10 m grids takes the form of one of the six land use categories classified in Section 5.3. Data on adjacent land use rely on the most frequent land use category out of eight grids (up, down, right, left, and four diagonal directions). Other factor values are represented by their respective attributive values of 100 m grids containing the 10 m grids. The operation that identified these attributive values is detailed in endnote 4. Figure 5.3 illustrates the values for each factor thus obtained for 10 m grids within the area under study.

Correlation analysis of these 11 factors shows that there are relatively significant correlations between "present state of land use" and "adjacent land use," between "distance from stations" and "time from the center of Tokyo," and between "land use zoning" and "floor-area ratio" ($R > 0.9$). In the light of this result, eight factors are selected for the land use estimation model, as underlined in Figure 5.3: present land use, distance from station, distance from trunk roads, change in population, altitude, gradient, land use zoning, and years after project completion.

5.4.2 Land Use Change Models by NN

The following two land use change models (illustrated in Figure 5.4) are examined in this study.

Total model: Land use is estimated from eight factors (including "present state" or land use in 1984) as input factors in one model.

Category model: Land use is estimated from the remaining seven factors as input factors in six models that belong to the six categories of land use in 1984.

What is used here is the back propagation (BP) model,[5] which is a type of NN. The BP model is composed of input layer, output layer, and hidden layer. Factor values (eight or seven, as shown in Figure 5.5) for each grid are input into the neuron of the input layer, while the neuron of the output layer outputs the estimated value of the six land categories (0.0~1.0 each). Land use with the highest value, among these six values, is decided as the estimated land use for that grid. A three-layer structure including input layer, one hidden layer, and output layer is used. The number of hidden neurons is set at six.[6] Each neuron in each layer is connected with all neurons of the adjacent layer, and those connections are respectively given the variable weight (w in Figure 5.4). The learning process is repeated by alternation of two operations: the forward operation (in which the value from the preceding neurons is changed by the weights and the changed value is sent to the following neurons) and the backward operation (in which the weights are changed to decrease errors between output value (estimated land use) and training value (actual land use)). The learning process is repeated until output value matches training value.

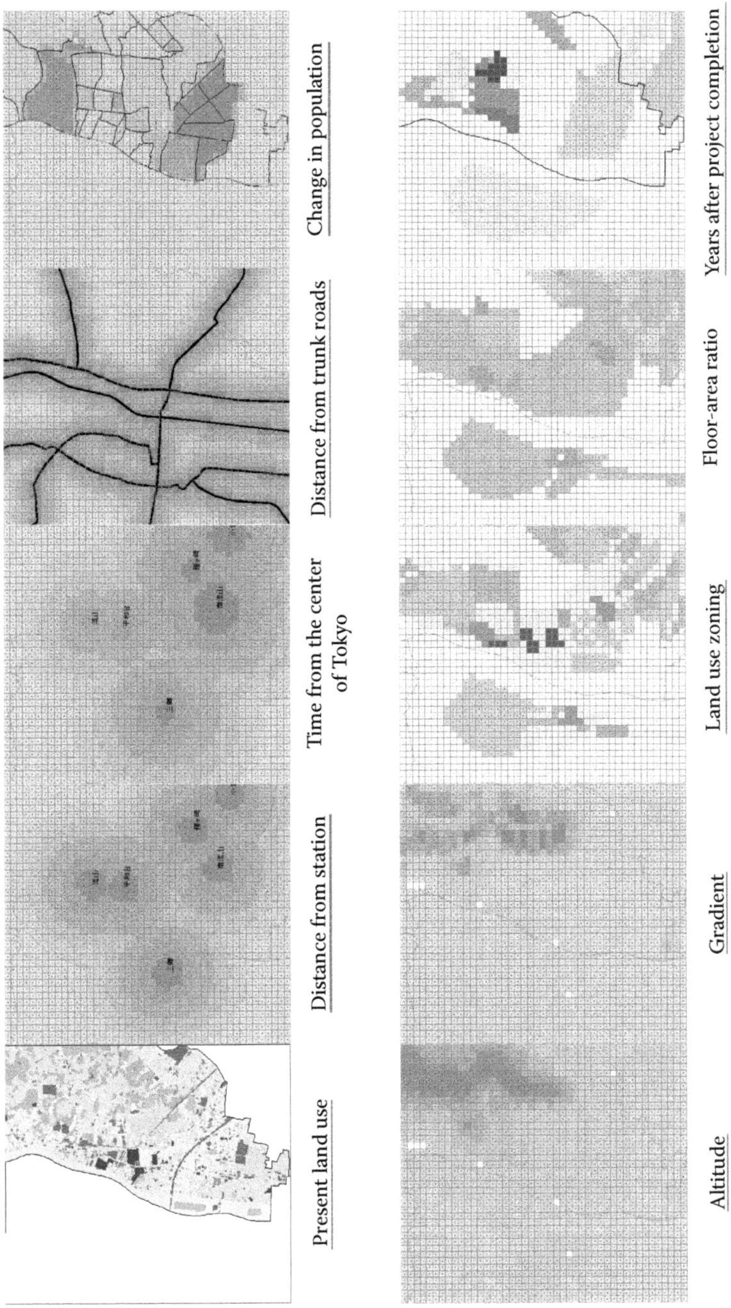

FIGURE 5.3 Factors causing changes in land use.

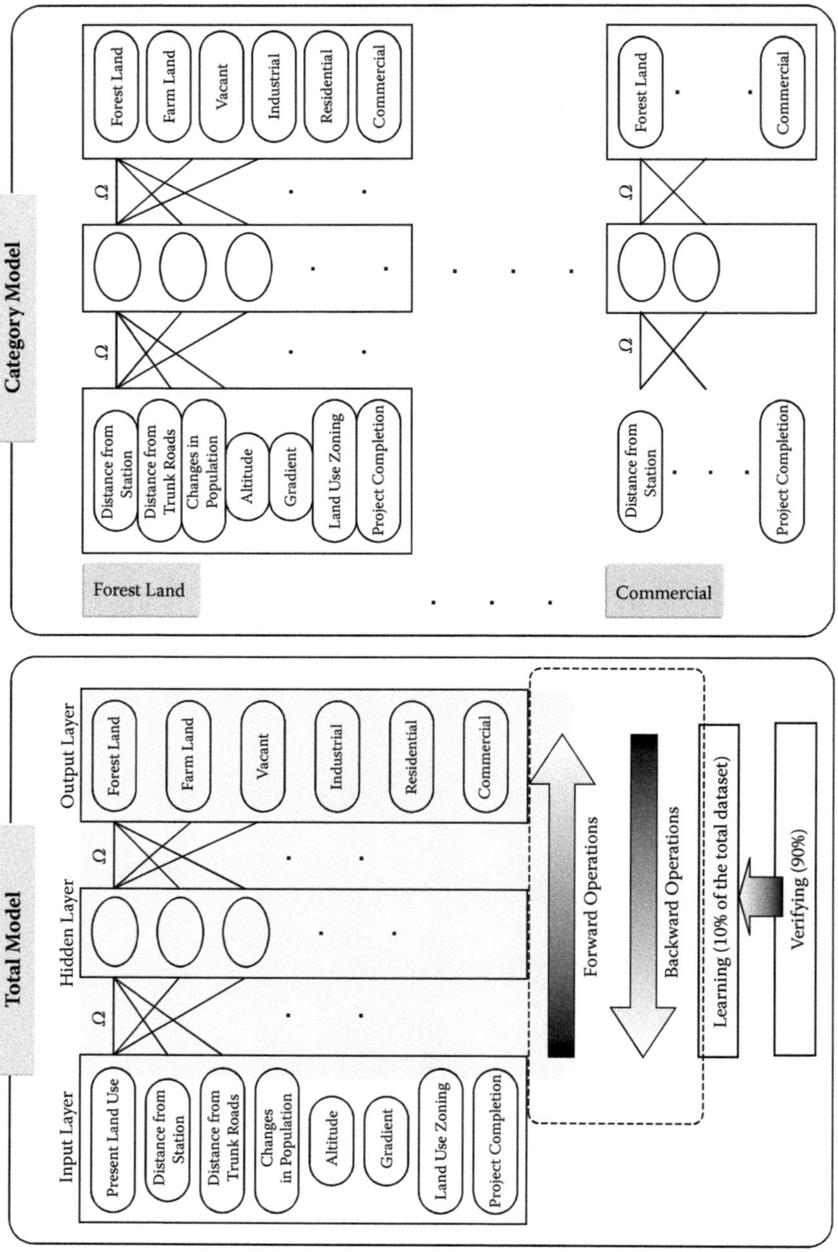

FIGURE 5.4 NN model estimating land use.

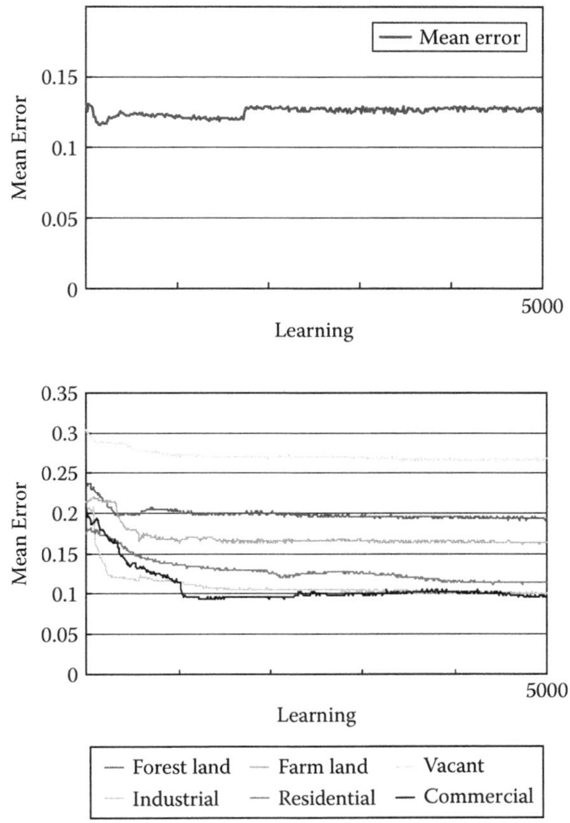

FIGURE 5.5 Mean errors between output and training value.

It should be noted here that out of 85,281 grids within the study area, the number of grids belonging to six categories of land use (all data to be analyzed in this study) totals 59,677. All of these make their respective datasets, with their factor data attached. From them, 6,000 datasets, or about 10% of the total, are randomly sampled as NN learning data, and the remaining 53,677 datasets, or about 90%, are used to verify the fitness of the models. The learning process was repeated 5,000 times before no further decrease in error between output value and training value was observed.

5.4.3 Analysis of Learning Result

Figure 5.5 shows a comparison between the two models made after 5,000 times of learning. The figure shows that the errors between output value and training value have gradually decreased in each model, and that learning by the category model has progressed more smoothly than that by the total model.

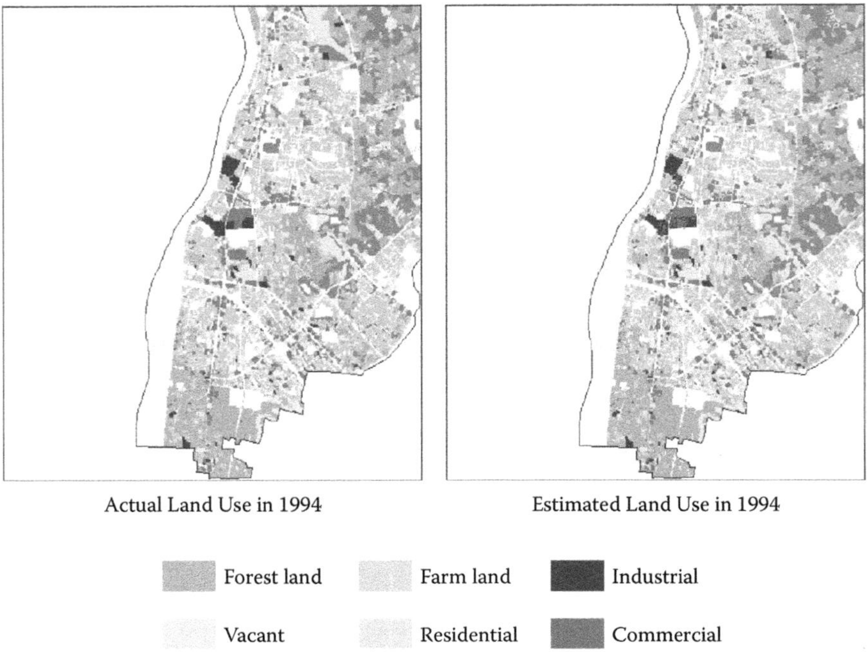

FIGURE 5.6 Actual and estimated land use.

This was followed by a test of the effect of learning by inputting the remaining 90% of data into the developed NN model. Comparison of the estimated land use output by the model with the actual land use gave the correct answer ratio of 67.1% for the total model and 74.2% for the category model (broken down by category: 78.6% for forest, 78.3% for farmland, 45.3% for vacant, 79.9% for industrial, 80.7% for residential, and 87.0% for commercial). In each of the two models, the null hypothesis of the learning effect was rejected by the multinomial test at a significance level of 1%, and thus the effect of learning is confirmed.

We chose the category model that had shown a smooth advance of learning and a high ratio of correct answers, and made it our land use estimation model. Figure 5.6 illustrates the land use in 1994 estimated by this estimation model and the actual land use in that year. The reproducibility ratio was 73.7% for all grids in the study area, including those of both learning and verification datasets. Goodman–Kruskal's correlation coefficient gained from the table (the actual land use by the estimated land use in six categories) shows a high correlation of $\gamma = 0.78$. Thus, this model has proved itself a highly applicable model that reflects the reality.

Correct answer ratios for all categories show that the ratios for all categories except for vacant lot are close to 80%, while the ratio for the latter is under 50%. This suggests that unlike those five categories, the vacant lot land use is on the way to a certain form of change. Its ratio of correct answers can probably be raised by special training data in characteristic categories (e.g., land use before the land became vacant), rather than random training data.

5.5 QUANTITATIVE ANALYSIS OF FACTOR EFFECTS ON LAND USE CHANGES

5.5.1 Definition and Calculation of Factor Effects

As the model developed in the preceding section shows a high reproducibility, the degree of factor effects in this model seems to represent the factor effect of actual changes in land use. This section tries to compute the effects of the factors by analyzing the inner structure of the model.

Factor effects numerically indicate the effects of seven factors for changes that were input in land use change models (distance from station, distance from trunk roads, change in population, altitude, gradient, land use zoning, years after completion of the land readjustment project) on the output values of their respective land use types (forest land, farmland, vacant, industrial, residential, commercial). Factor effects depend on the causality index by Enbutsu et al. (1990).[7] The positive value of the index indicates that the quantity of change in land use tends to increase when input factor value increases, while the negative value indicates that the quantity of change tends to decrease. It has also been found that the larger the absolute value of the index, the heavier the effects.

Figure 5.7 shows the results of factor effects. Change from a certain land use category to another tends to occur at the grids, which have some conducive factor conditions. The following are seven typical cases and their conditions:

1. Forest land to vacant: Close to station, far from trunk roads, little change in population
2. Forest land to residential: Close to station, increasing population, fewer years after readjustment project completion
3. Farmland to vacant: High altitude, slight gradient
4. Farmland to residential: Close to station, high altitude, slight gradient
5. Vacant to residential: Far from station, slight gradient, fewer years after project completion
6. Vacant to commercial: Close to trunk roads
7. Commercial to vacant: Far from station, fewer years after project completion

5.5.2 Analysis

Observing the effects of land use changes by category (Figure 5.7), we can see under what conditions a certain category of land use is likely to occur. A change into residential is greatly influenced by distance from the station, change in population, gradient, and years after project completion, while a conversion into commercial land is affected heavily by distance from trunk roads. These results show that when a certain place (10 m square grid) has specific conditions relating to these factors, we can predict a land use change in that place fairly well.

This study proceeds by examining the total effects of each factor on land use as whole categories and the comprehensive effect of each factor, by computing the ratio of the comprehensive effects by factors to the sum of the absolute values of the

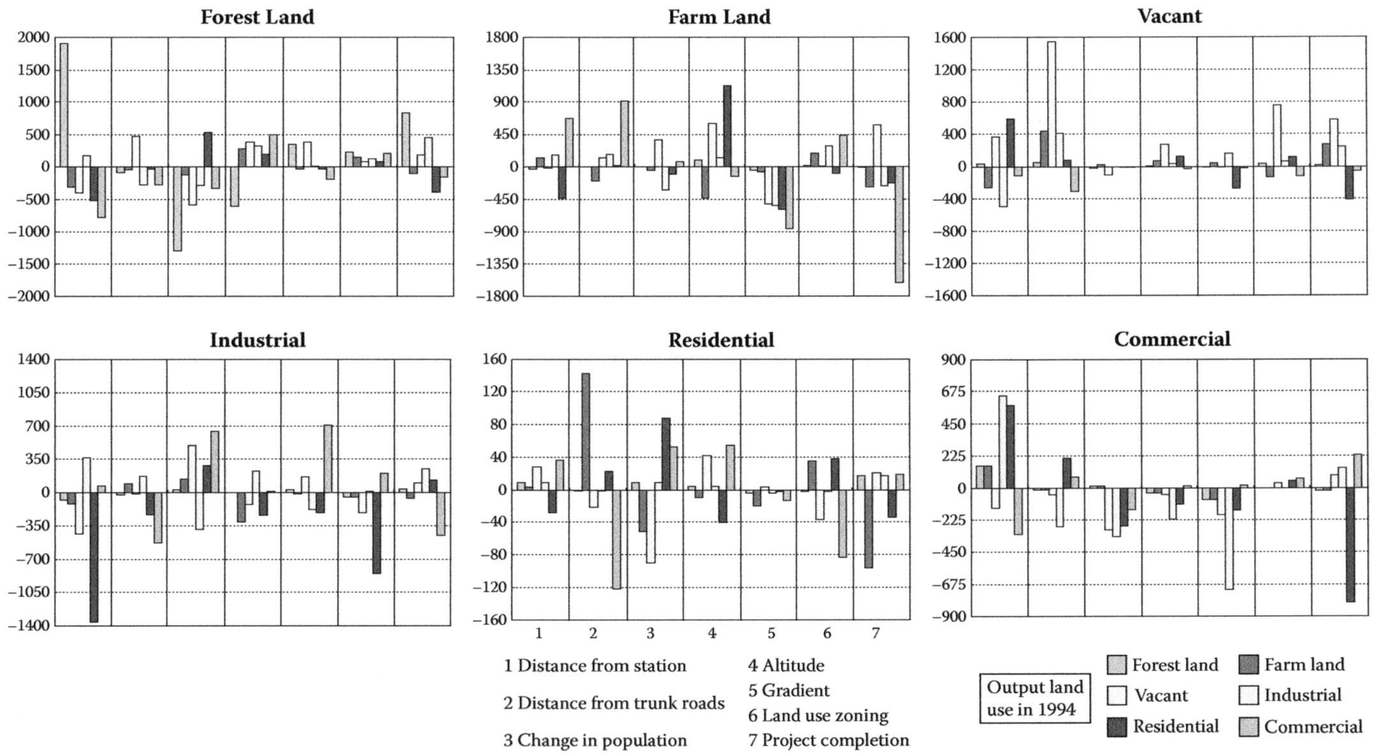

FIGURE 5.7 Effects of factors.

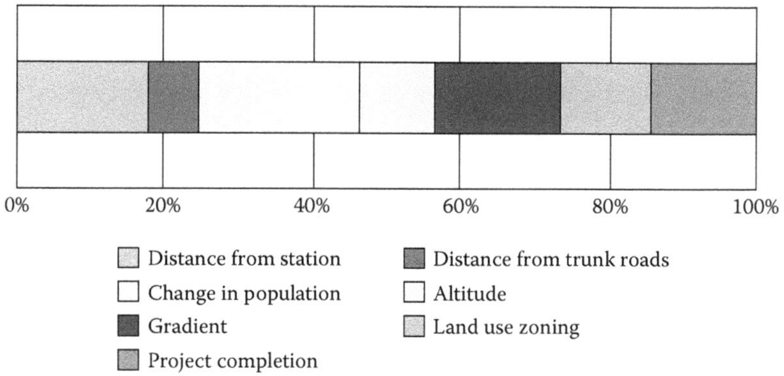

FIGURE 5.8 Comprehensive effect of factors.

results gained in the preceding paragraph. Figure 5.8 shows that the comprehensive effect is strong in the order of change in population, distance from station, gradient, years after project completion, land use zoning, altitude, and distance from trunk roads. This order applies to the degree of effects of these factors on land use change. In this list, "change in population" comes first, as seen in the findings of prior studies, but not a single factor stands out. All factors have similar ratios, as the figure shows. The fact that "years after project completion," a factor adopted in this study, has a stronger effect than "land use zoning" suggests that an impact on land use transition is greater from drastic structural change, such as an urban project, than from static land use regulation, such as land use zoning. Another fact, that out of the natural factors, "gradient" has a greater effect than "altitude," reflects the actual process of land development. All this suggests that a land use estimation model should be composed of many of these factors.

5.6 CONCLUSIONS

Our research was about land use changes that are taking place in the northwestern part of Nagareyama City, 30 km from the center of Tokyo. It aimed at (1) quantitative determination of those changes by category of land use, (2) modeling the structure of land use changes by use of NN, and (3) analyzing the degree of the effects of the factors for land use change. Conclusions are as follows:

1. The ratio of unchanged land use in the area under study is high in the order of residential and commercial land. The ratio of changed land use is high in the order of vacant, industrial, and forest land. The conversion takes place from forest land and farmland to vacant lot, and further to residential section. The tendency of the land use change is toward expanding residential use area.
2. We developed an NN model to estimate land use after 10 years of change by six categories of land use. The model has proved fairly accurate and applicable. The matching rate has reached over 70% of about 60,000 ten meter grids in the study area.

3. As regards the factors causing land use changes, "distance from the station," "change in population," "gradient," and "years after completion of readjustment project" have a stronger effect on conversion into residential land. "Distance from trunk roads" has a stronger effect on conversion into commercial land.
4. The comprehensive effect on land use change is strong in the order of "change in population," "distance from the station," "gradient," and "years after projection completion." There is little significant difference between those factors, and this suggests the need to employ many factors in forming the land use estimation model.

With the opening of the Tsukuba Express in 2005, there have been notable changes in four factors: distance from station, change in population, years after project completion, and land use zoning. If newly gained data on these factors are employed in the models, they should engender more accurate estimation of future land use change.

The model has some issues for further study. It can be more generally applicable if the area for gaining training data grids is widened and factors of regional characteristics are employed.

ACKNOWLEDGMENTS

The authors thank Nagareyama City for permitting us to use relevant data and materials. The authors also thank Atsuyuki Okabe and Yasushi Asami for their valuable comments.

SUPPLEMENTARY NOTES

1. This is a translation of the joint study by Ito and Murata (2000), with minor modifications to improve presentation.
2. For example, land use in residential and commercial areas changes in line with the purpose of the land use zoning. Furthermore, the suburban areas along bypasses and beltways present a different trend of change. Such multiple factors should be taken into account.
3. Land use changes at two target points in time, which were aggregated according to land use subclassification, indicate that the ratio of unchanged land use grids (e.g., A in terms of land use in 1984 was still A in 1994) exceeded 95% in all these areas.
4. Factors are defined as follows: distance from the station = at the 250 m interval for the distance from the centroid of 10 m grids to the station point; distance from the center of Tokyo = at the interval of 5 minutes from the shortest time (35 minutes between the grid centroid and Tokyo's Yamanote Line via a nearby station); distance from trunk roads = at a 100 m interval from the grid centroid to the prefectural road or main local road; changes in population = at the interval of 10 persons/ha for changes in population density in 1984 to 1994 based on data by town; altitude = at a 3 m interval; gradient = at a 7% interval; years after readjustment project completion = at a 5-year interval. For land use zoning and floor-area ratio, the urban planning map of 1983 has been consulted.
5. NN is an artificial model of the working of the cranial nerve system, and comes in two types. The Hopfield model imitates equi-directional thinking, like association of ideas, and the back propagation model imitates think flow in one direction, like conditional judgment. This study uses the latter in order to output land use conclusively from factor conditions.

6. In the cranial nerve system, the number of hidden layers is believed to be larger, but NN models often adopt one layer in order to hold down the number of variables. The number of hidden neurons is usually set to about the same as that of output neurons. After trying six and seven hidden neurons, this study chooses six, a number that provides a higher ratio of correct answers.
7. Causality index $C(y, b; x, a)$ indicates the effect that is received when input value is $x = x_a$ so that output value comes to $y = y_b$. It depends on the sum of the effects of all courses from input x_a through all neurons in the hidden layer to reach output y_b. The effect that passes the hidden neuron j is expressed as

$$[\partial y_b / \partial x_a]_j = f(u_b) w_{bj} f(u_b) w_{ja}$$

where f is the conversion function, u is the product summation, and w is the weight parameter.

Therefore, causality index is defined as

$$C(y, b; x, a) = \Sigma_j [\partial y_b / \partial x_a] = \Sigma_j f(u_b) w_{bj} f(u_b) w_{ja}$$

REFERENCES

Arai, T., Minemura, M., and Akiyama, T. 1999. Simulation of urban land-use dynamics with a grid-based model: An application of cellular automata. *Papers and Proceedings of the Geographic Information System Association* 8:69–72.

Chishaki, T., Noda, K., Konaga, D., and Tatsumi, H. 1990. Relationship between land-use and specified land use zones—Case study in Fukuoka City. *City Planning Review* 195:56–64.

Edamura, T., and Kawai, T. 1992. Development of micro-land-use model by neural network. *Papers on City Planning* 27:175–80.

Enbutsu, I., Baba, K., Yoda, M., and Hara, N. 1990. Extraction of explicit knowledge from an artificial neural network. *IEEE Tokyo Section* 30:101–4.

Ito, F., and Murata, A. 2000. Artificial neural network model estimating land-use: Change in Nagareyama City. *Papers on City Planning* 35:1129–34.

Kawakami, Y., and Honda, Y. 1994. Analysis of the relationship between zoning ordinance and land use transition in local city—A case study of Fukui City. *Papers on City Planning* 29:475–80.

Osaragi, T., and Masuda, K. 1995. Improvement of Marcov chain model for land-use forecasting. *Papers and Proceedings of the Geographic Information System Association* 4:71–74.

Takizawa, A., Kawamura, H., and Tani, A. 1997. Formation of urban land use pattern by genetic algorithm. *Journal of Architecture, Planning and Environmental Engineering* 495:281–87.

Yoshikawa, T., Okabe, A., Asami, Y., and Kaneko, T. 1990. An analysis of land use transition in view of zoning regulation. *Papers on City Planning* 25:373–78.

Section II

Urban Analysis in the Social Sciences

Toru Ishikawa

The second section of this volume is a compilation of chapters on urban analysis in social science, illustrating close relationships between the two fields of research. One of the most closely related social science disciplines to urban analysis, particularly from a theoretical point of view, is urban economics or regional science, and there is a body of literature on housing market analysis, especially through normative approaches such as utility maximization or location optimization models. In addition to economic analysis of this sort, consideration of regulation issues is also important in urban housing research, particularly for the planning of good residential environments. Kume's chapter is a good example of this line of research.

Kume examines the Japanese rental housing market, by looking at rental housing law systems such as the just cause system and the fixed-term rental system. The just cause clause, introduced in 1941, protected existing renters, but at the same time caused an increase in house rent and poor living environments. For example, in Tokyo there are areas where decayed rental apartments are concentrated, which poses a problem of poor resistance to fires or earthquakes. To cope with these problems, the fixed-term rental system was introduced in 2000; however, the system is not applied to existing contracts, and as Kume argues, it is difficult for landlords to objectively judge whether the rejection of lease renewals at the time of term expiration warrants sufficient just cause. Kume overviews the history of the just cause system and examines current rental systems in England, Germany, and France. Based on an analysis of judicial precedent data on just cause for land and house leases, he offers suggestions about a legal revision of the just cause system. This chapter shows how research in urban analysis contributes to the field of urban and housing planning through economic and judicial analysis.

Another social science discipline that is closely related to urban analysis is geography. This is especially true for spatial-scientific studies of spatial distributions and urban structures, prompted by the quantitative (or theoretical) revolution in the late 1950s and early 1960s. The major focus of these studies is spatial *form*, and a wealth of research findings has been obtained about urban structure, or *where* things and people are. Masuda's chapter showcases contributions by research in urban analysis to the understanding of the structures of Japanese cities.

Masuda's chapter compared employee distributions of 65 medium-sized cities and population distributions of 245 cities in Japan. To do that, Masuda uses a graph theoretic method proposed by Okabe, which extends Clark's classic work of one-dimensional descriptions of population density to a two-dimensional analysis of distributions, by qualitatively describing global structures of density surfaces in terms of peaks, bottoms, and cols. Masuda shows differences between employee and population distributions, and discusses characteristics of the spatial patterns of city centers and subcenters.

As well as spatial form, research in urban analysis looks at *processes* underlying various spatial phenomena, as another important research theme. That is, urban analysis researchers aim to answer *how* and *why* things or people are where they are, and to do that they look into cognitive processes underlying human spatial behavior. There are many studies of consumer behavior, particularly by applying spatial interaction models such as the gravity model, Huff's model, or various other potential models. And with the emergence of behavioral approaches in geography, cognitive analysis of consumer behavior attracted many researchers' interest, for example, Rushton's analysis of spatial behavior by revealed preferences, and Downs' study of subjective images of shopping centers.

Yamada and Sadahiro's chapter examines consumers' evaluation of department stores on the basis of survey-questionnaire data, and discusses relationships between characteristics of department stores, consumers' evaluation of the stores, and their shopping behavior. They report a positive effect of the frequency of consumers' visits to a particular department store on their evaluation of that store. This chapter exemplifies how an analysis of people's perception or cognition contributes to an understanding of their behavior, in this case shopping behavior. Such a cognitive analysis also has potential to provide insights into a disaggregate analysis of people's spatial behavior.

Ishikawa et al.'s chapter is an environmental psychology study, and it examines how people in urban space perceive spaciousness, by conducting a controlled experiment in a computer-simulated 3D virtual environment with human participants. Ishikawa et al. apply the concept of *isovist* which was proposed by Benedikt and Burnham to analyze static space, to an analysis of dynamic space, and discuss how the location of an open space on a block affects its perceived area. This chapter shows how findings from a cognitive study of the interaction between humans and the physical environment can be applied to urban planning, for example, the design and construction of open spaces that look larger with a fixed area under budget constraints.

As the reader may notice, the overarching theme for all chapters in this section, and in this volume is *space*. That is, urban analysis aims to study scientifically spatial phenomena on the basis of theoretical models and empirical data, and thus it can now be regarded as an important contributor to the field of spatial information science. Hopefully these chapters help the reader to learn theories and methods of urban analysis and to find interesting research questions that deserve further investigation.

6 Empirical Analysis of the Evaluation of Judicial Precedents of Compensation Fees for the Surrendering of Lease Premises

Yoshiaki Kume

6.1 INTRODUCTION

The powerful limitations placed on lease terminations and the renewal rent regulation under the just cause clause (House Lease Law, Article 1.2), introduced under the 1941 Amended House Lease Law, obstructed the realization of high housing standards in the post-war metropolitan housing market. Specifically, although protecting existing renters, they made rent expensive and the living environment poor for new renters, who were the source of potential demand. They also caused extreme shortage of rentals with large space and a large number of rooms, as well as other problems, including the expansion of the studio rental market, which has high renter turnover.

In such an environment, the fixed-term rental system was created under the Amended Land Lease and House Lease Law implemented in March 2000.* Under this law revision, however, (1) the fixed-term lease contracts were introduced only as an option for new contracts and were not applied to existing contracts (so-called vacancy decontrol), and (2) existing ordinary lease contracts under the protection of the just cause system that were in existence at the time of implementation of the amended law were not allowed to be converted to fixed-term lease contracts, even at the time of contract renewal.

Therefore, with respect to existing lease contracts, the decision as to whether or not there is sufficient just cause in the event that a landlord refuses lease renewal

* Fixed-term rental contracts have steadily spread. For example, in rentals targeting families with floor space of more than 70 m^2 in the 23 Tokyo wards, these contracts have been used in about 5% of multi-housing units and in about 25% of single-housing units. Please refer to Kume (2003).

at the time of contract term expiration continues to be made by comparing and measuring the interests of the landlord versus the interests of the renter. Even if a property payment, or a compensation fee for surrendering the premises, is required, since this is ultimately decided on the basis of the judge's viewpoint, it is extremely difficult to estimate in advance. Consequently, landlords cannot make business plans for rebuilding decayed buildings or for making repairs to improve the fire or earthquake resistance of their buildings. In central areas of the city with prime real estate, this leaves a concentrated housing district with its foundations untended. The buildings are abandoned as a negative legacy of Japan's post-war high growth from the perspective of both the improvement of housing standards and the risk of disasters.

In terms of legislative theory, just as in European countries, a type of contract should exist as one type of lease contract that imposes a limitation on terminations with the condition that renewal rent be revised in line with market rent. The requirements that constitute just cause, however, which is necessary for the rejection of lease renewals by landlords at term expiration, should be strictly objective and possible to estimate. Otherwise, the uncertainty of running a rental property will become too great for the landlord and, as a consequence, societal loss, meaning the ineffective use of scarce resources such as land and buildings, will increase.

This chapter looks back once again at the history of the creation of the just cause system and reexamines its legislative intent, among other factors. In addition, we will analyze the current rental systems in England, Germany, and France, where a contract type exists that includes a limitation on lease termination under the just cause system, and we will illustrate the three points that differ from the Japanese legal system: (1) the requirements to satisfy just cause are listed specifically on a limited basis as law and only the circumstances of the landlord are considered; (2) renewal rent regulations exist in such a way that renewal rent is revised in line with market rent; and (3) the paying or receiving of a compensation fee for the surrendering of the premises does not exist under law or custom. We will also quantitatively analyze 176 judicial precedents pertaining to just cause between April 1980 and March 2003, and offer suggestions on the legal revision of the just cause system after analyzing the requirements needed to satisfy just cause, and particularly the standards of evaluating compensation fees for the surrender of premises under the decision standards of current courts.

6.2 LAW PERTAINING TO THE RENTER PROTECTION SYSTEM AND ECONOMIC ANALYSIS

6.2.1 History of the Introduction of the Renter Protection System

6.2.1.1 Law Pertaining to Building Protection (1909)

Originally, provisions concerning the lease of land or buildings were stipulated under Civil Code, Section 7 Lease, Articles 601 to 622. The special law under the Civil Code that was stipulated concerning the lease of land or buildings in addition to the above is the Law Pertaining to Building Protection created in 1909 (May 1, 1909, Law No. 40).

Under the basic principle of the Civil Code, the requirement to duly assert against third parties with respect to change in control of property for real estate is the registration. Even if the party that owned a building based on surface rights or right to lease had registered the ownership or lease of the building, however, if a registration of the land's surface rights or right to lease had not been made, the party would not be able, for example, to duly assert against a third party to which the land was assigned. Moreover, depending on the intention of the landowner, the party may have no choice but to tear down the building. Therefore, Law Pertaining to Building Protection, Article 1 stipulated that "in the event surface rights or right to lease the land exist for the purpose of owning a building, and the owner of such surface rights or the land lessee owns a registered building located on such land, such surface rights or land lease shall be sufficient to duly assert against a third party even if they have not been registered," and thus protected the continued existence of buildings that are expensive investment assets.

6.2.1.2 House Lease Law (1921)

The law that carried through this legislative intent and enhanced the rights of land lessees so as to further protect the continued existence of buildings is the Land Lease Law (April 8, 1921, Law No. 49) created in 1921. Article 2 of the Land Lease Law stipulated that "the continuation period of a right to lease land whose purpose is to own sturdy buildings made of stone, earth, brick, or a similar structure shall be 60 years, and the continuation period of a right to lease land whose purpose is to own other buildings shall be 30 years," and stipulated the minimum continuation period, etc., of surface rights and rights to lease land.

The House Lease Law (April 8, 1921, Law No. 50) was created at the same time as the Land Lease Law. Lack of housing was a common phenomenon in the metropolitan area, and since under the Civil Code principle, landlords could terminate leases that did not have term stipulations by providing 3 months' advance notice, there were circumstances in which renters were forced to surrender the premises after a short period and had difficulty finding alternative housing. From this standpoint, the House Lease Law stipulated the statutory renewal relating to lease contracts. The specific details are as follows:

1. With respect to leases with a stipulated period, in the event that the parties do not provide notice of, for example, renewal refusal at least "6 months prior to and within 1 year" of term expiration ("notice of renewal refusal or, in the event conditions will not be changed, notice of non-renewal"), the contract will be renewed under the same terms and conditions.
2. Even if a notice of renewal refusal, etc., is given, in the event that the renter continues use of premises even after term expiration and the landlord does not provide an objection without delay, the lease will be renewed under law (Clause 2 of the same article).

6.2.1.3 Rent Control Act (1939)

In 1937, as the Second Sino-Japanese War erupted and Japan prepared for war, there was a tendency for the population to concentrate in metropolitan areas, and rent skyrocketed. As a countermeasure to this situation, in 1939 the Rent Control Act (1939 Imperial Ordinance No. 704) was enacted under the National General Mobilization Law. The Rent Control Act maintained the maximum rent amount (1) for renewal contract properties from August 4, 1938, and before, the same level as at such date (Rent Control Act, Article 3, No. 1), and (2) for contract properties after such time, at the same level as at the time of contract execution (Rent Control Act, Article 3, No. 2).

6.2.1.4 Introduction of the Just Cause System (1941)

Rent control without limitations on lease terminations has, however, no effect. In Europe and the United States, rent control was also mandated to protect renters at times of insufficient resources before and after World War II, but these always included limitations on lease terminations. In reality, in many cases, landlords who wished for income beyond the mandated rent refused renewals and demanded surrender of premises from renters. As a countermeasure to such situations, in 1941 under the Amended House Lease Law the provision for just cause that was needed for renewal refusal or request for termination was created. Specifically, as in House Lease Law, Article 1.2, a limitation on lease terminations was introduced, stipulating that "in the event the lessor of the building requires the use of such building for himself, unless the lessor has other just cause(s), the lessor may not refuse lease renewal or request lease termination."

Under the literal interpretation, as long as the lessor of the land/building required the use of such land/building for himself, that in itself would constitute a reason for renewal refusal and request for termination, etc. This being the case, the judicial precedents from such time also held the same thinking (Supreme Court Ruling 1943/2/2, Civil Case Judicial Precedents 22/57).

Amidst housing difficulties from war damage and chronic housing shortages owing to flow of population to major cities as a result of post-war high economic growth, gradually judicial precedents accumulated that interpreted just cause more narrowly and strictly. In particular, it reached a point where not only the interests of both the landlord and the renter were to be compared and measured, but other various public interests and societal factors were also to be considered, and landlords could not use the mere requirement to use the building for their own use as just cause (Supreme Court Ruling 1944/9/18, Newspaper 717/14; Supreme Court Ruling 1954/1/22, Civil Case Judicial Precedents 8/1/207; etc.).

6.2.1.5 From Rent Increase/Decrease Demand Rights to Renewal Rent Control

In addition, as soon as the limitation on lease terminations called just cause was introduced, the rent increase/decrease demand rights functioned as renewal rent control.

House Lease Law, Article 7 stipulates that the parties may demand increase/decrease of rent in the event that the rent becomes inappropriate owing to rise/fall of land/building prices, other economic factors, or compared with surrounding buildings. As post-war urbanization progressed and inflation occurred, market rent consistently went up, and there were many disputes surrounding rent revisions at the time of contract renewals. In these instances, it became commonplace for courts to decide the renewal rents at levels that were significantly lower than market rent.

6.2.2 Interpretation Theory of Legislative Intent by Civil Law Specialists and Its Reevaluation

Let us now examine how civil law researchers interpreted the legislative intent of the statutory renewal system and the just cause system. We will look back on the history of the overall academic theory and reevaluate these systems from the perspective of today's laws and economics.

6.2.2.1 Legislative Intent of the Statutory Renewal System

6.2.2.1.1 Social Legislation Policy Theory by Civil Law Researchers

With respect to the creation of the statutory renewal system under the House Lease Law (1921), the responsible persons for the legislation considered it as legislation for social policy.* Suzuki (1959) also understood those times to signify

> when housing difficulty in the city worsened significantly, and as a result, housing disputes were frequent and there were active movements by renter and landlord associations. In addition, labor disputes and tenant farmer disputes became frequent, unemployment increased due to the financial crisis brought on by the Rice Riots and the end of the war, and conflict between the social classes heightened ... social policy ideas began appearing.... Farm Tenancy Arbitration Act (1924) and Labor Dispute Arbitration Act were in fact enacted. With respect to the housing problem, as part of the aforementioned trend, the Land Lease Law and the House Lease Law were implemented.... [I]t was common knowledge for people that the Land Lease Law and the House Lease Law were enacted as part of a social policy. (pp. 11–12)

6.2.2.1.2 Statutory Renewal System for Reduction of Transaction Costs

The statutory renewal system, however, should be understood as a system for allocating resources efficiently and not as a system for equal distribution or social policy legislation. From this perspective this policy can be rationalized.

In a lease, even under a short-term contract, it is natural for the landlord and the renter to expect that they will repeatedly renew the contract and for the contract relationship to continue. Even for the landlord, he cannot ignore the fees necessary for searching for a new renter. In closing a lease contract, information is not known equally between the two parties, since the renter's character and personality cannot be known sufficiently. For example, at the expiration of the contract term, such as after 2 years, it would be natural for the landlord to wish that the existing renter would renew the contract and continue living in the premises if such renter had treated the premises with care, did not cause disturbance to the neighbors, and had not delayed paying the rent.

Similarly, for the renter, the cost for searching for a new rental property is also not cheap. Moreover, by continuing to live in the same area and housing unit, typically the renter will have become familiar with and attached to the place.

* Government committee member Kisaburo Suzuki stated in the 44th Imperial Diet House of Peers Special Committee for a Case Outside the Land Lease Law Proposal, "No matter how much the renters, who are the weak, suffer, if the landlords can do anything they wish under the principle of freedom of contracts ... we must think of this carefully as a social policy ... we must protect the weak ... considering their standpoint" (stenographic records of minutes no. 2, p. 14).

In such an environment, if the renter receives a request from the landlord to surrender the premises owing to lease contract termination 1 day prior to contract term expiration, not only will the renter be perplexed but, if the handover period is short, he will also incur significant costs in searching for a new rental property. The implementation of the statutory renewal system prevents such transaction costs from becoming extremely high.*

6.2.2.2 Legislative Intent of the Just Cause System

6.2.2.2.1 The Effective Use of Resources Theory, the Modernization Legislation Theory, and the Social Policy Legislation Theory by Civil Law Researchers

It is evident that after the war, when just cause was starting to be interpreted narrowly and strictly, most business and commentary publications periodically published by business people during this period represented just cause as "that whose purpose is to maintain the reasonable use of property and to equally distribute housing" (Suzuki, 1959, p. 21).

For example, Hirose (1950, p. 191) stated: "The purpose of this Article is to maintain in the most reasonable way the legal relationship of property use which is social and economic." Furuyama (1951, p. 11) also stated that "in today's world … we cannot employ the 'Principle of Renter Protection' as is without any conditions. Rather the legal importance of the relationship of real estate use is moving towards the point of 'how can both parties contribute to the most effective use of national resources.'"

The legislation theory presented by Watanabe (1951) relating to the characteristics and provisions of the House Lease Law is considered as House Lease Law = Modernization Legislation Theory in the history of academic theory. For example, Watanabe (1951, p. 24) analyzed that "the basic form of lease relationships after the Meiji Era is a reflection of the human and physical relationships that existed between the tenant and the landlord in ordinary life during the Tokugawa Era … the tenant had a slave-like role to the landlord, and it was a relationship in which the landlord was dominant," explaining the feudalistic characteristic that remains in Japanese lease contracts. He also stated that the House Lease Law enacted in 1921 "could not touch the roots of the quasi-feudalistic structure of lease contracts even though it advocated renter protection legislation, which shows its limitation as social policy legislation." He then commented on the House Lease Law amendment in 1941 that "the large-scale limitation on the termination request rights of the landlord helped to promote the breakdown of previously quasi-feudalistic lease relationships."

Suzuki (1959, p. 37) critically analyzed the aforementioned research and pinpointed the House Lease Law as the "social legislation that uses the medium of contractual relationships and is based on the sacrifices of landlords." In other words, Suzuki (1951, p. 49) regarded the House Lease Law as "that which pursues the security of housing for the general public and guarantees a humane life." His understanding was that "the

* The fact that this was made into a mandatory provision was against the principle of freedom of contracts. However, even for the landlord, it is rare that a situation randomly occurs that would make the landlord not wish to renew the contract. There is little loss to the landlord by this being a mandatory provision.

correct housing policy is not to delegate the duty of providing secure housing for the general public to private landlords, but to eliminate housing difficulties through the construction of public housing through national budgeting," meaning that the protection of the weak through redistribution is the role of the government. He stated, however, that "considering the current housing situation and the status of the national budget amount allocated to the resolution of housing difficulties (especially for the construction of public housing), the overall aim of the House Lease Law contract termination restriction should be considered as something positive."

6.2.2.2.2 *Strictly a Complementary Method for Rent Control and Securing Its Effectiveness*

The implementation of rent control under the Rent Control Act (1939) is understood as a price control act aiming to avoid deterioration of public peace and social unrest from riots, etc., owing to price rises stemming from shortages of living supplies, while the country needed to invest significant resources such as capital and labor into the execution of the national aim of war. Therefore, it is not a policy for equal distribution but a strategy for effectively allocating resources. The just cause system (1941) is also strictly a complementary method for securing the effectiveness of rent control. At the time of implementing the system, there was no legislative intent to protect the weak through redistribution.

6.2.2.2.3 *Intent of the Just Cause System That Became Strict Owing to Judicial Precedents*

How should we interpret the intent of the just cause system that became strict owing to judicial precedents after the war? The Effective Use of Resources Theory by Hirose (1950) and Furuyama (1951) is not appropriate. If we were to focus on specific parties in a contract, a specific landlord and a renter, and the specific rental under such a contract, there may be an effect that prevents the idle use of resources that stems from the landlord's unwillingness to lease. However, in order to discuss the effect of law on the effective use of resources, one must analyze the entire market. As discussed later, the just cause system works in the opposite direction from the effective use of resources.

The Modernization Legislation Theory by Watanabe (1951) and others also does not provide sufficient basis to justify the just cause system that violates the major principle of the Modern Contract Law stipulating that "promises shall be kept."

It is true that during the post-war recovery and economic growth after World War II, a large population flow to the city continued and market land prices and rent skyrocketed, exceeding the general price rise of goods. Owing, however, to the Rent Control Act that remained for a certain period after the war under the Potsdam Imperial Ordinance, a tenant gain was created for renters who lived in the same property for a certain period of time, and rental rights became an asset with high value. At this point, a judge who had to resolve the one-on-one dispute between the landlord and tenant had no choice but to measure comparatively the interests and losses of both parties in deciding whether to terminate or continue the rental contract. The renter whose renewal rent had been kept far lower than market rent may have neglected to put in the efforts to increase his income or he may have spent his money wastefully. If he were forced to surrender the premises in which he had lived for years, and assuming his ability to pay rent had not changed, he would have had to live in a property under much worse conditions and face hardships with respect to housing. It can be inferred that under such circumstances, the tendency to protect renters strengthened and a case law that forced landlords to pay compensation fees for the surrender of the premises (monetary payment) in order to make up for just cause was created. In this sense, the legislative intent of the case law was nothing but a "redistribution policy in order to protect the weak by sacrificing the landlords," as indicated by Suzuki (1959).

The reason renters faced hardships with respect to housing, however, was because an effective renewal rent control existed owing to the implementation of the just cause system. Without the just cause system, they would not have had housing hardships and the construction of a case law that protected renters would not have been necessary. In other words, if we were to design a system from scratch, even if the housing policy for the weak, such as public housing, were not sufficiently in place, the just cause system cannot be justified from the Social Policy Legislation Theory. In other words, even if the current housing policy for the weak is insufficient, completely eliminating the just cause system will not cause any problems whatsoever unless we retroactively apply it to existing contracts.

6.2.3 Legislation Theory of the House Lease Law by Economists and Its Reevaluation

6.2.3.1 Beginning of Legislation Theory by Economists

It is fair to say that Iwata (1976) conducted the first research that analyzed economically the effect the house lease system was having on the housing market. Iwata (1976, p. 129) analyzed the effect, shown in Figure 6.1, as follows. S0 is the rental supply curve, D0 is the rental demand curve, P0 is the equilibrium rent, and Q0 is the equilibrium volume of rentals, before the implementation of the renter protection system. Once the protection system is implemented, the supply curve shifts to the left to S1. The shift to the right of the demand curve is comparatively small. As a result, the equilibrium volume of rentals is reduced to Q1 and the equilibrium rent rises to P1. Since excess consumers will also decrease at this point, it is shown that "more powerful lease rights do not necessarily lead to profit for the landlords as a whole."*

* Based on this analysis, Iwata (1976, pp. 136–38) pointed out that the two mutually contradictory tasks of "not drying up the supply of rental land and housing" and "protecting the renter" cannot be achieved through the Land Lease and House Lease Law. First, he suggested that "the aim should be to create an environment in which the land owner himself has no choice but to use the land effectively rather than making the land lease rights into real rights. Enhancing the capital gains tax on land is one of the most effective policies to this end," and promoted the correction of the land taxation system. Second, he suggested the implementation of a rent assistance system that "supplements rent for families with income that is lower than a particular level referring to market rent as the base and taking into account the family structure." He also insisted that the protection of lease rights be loosened concurrently, stating that "the just cause necessary to reject lease renewals should be interpreted more moderately than has been seen in post-war judicial precedents."

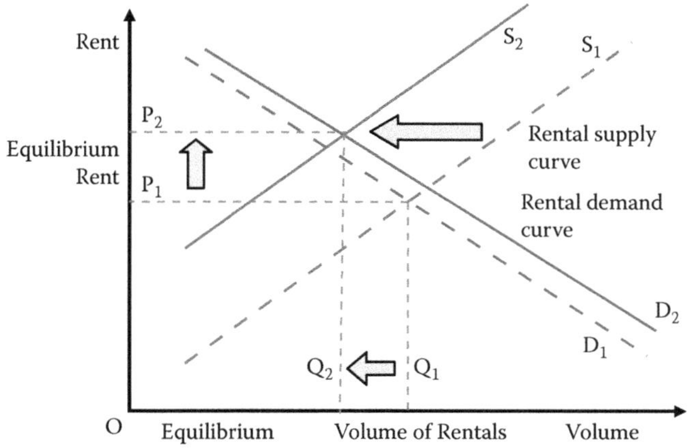

FIGURE 6.1 Effect of implementation of the renter protection system on the market (simple explanation).

6.2.3.2 Why Does the Equilibrium Volume of Rentals Decrease?

With respect to the fact that the leftward shift of the supply curve is much larger than the rightward shift of the demand curve, Iwata (1976, p. 129) only said that the reason was because "almost all renters are at such low income levels that they are unable to pay the price of lease rights turning into real rights," which is not sufficient explanation.

Kanemoto (1992), Iwata (1994), and Fukushima (1998), who analyzed this point in detail, stated that two conditions would make the leftward shift of the supply curve equal to the rightward shift of the demand curve: (1) information asymmetry between the landlord and the renter does not exist, and the level of renewal rent restriction and the amount of compensation fee to be paid for the surrendering of the premises can be estimated with certainty; and (2) the renter can borrow funds freely, meaning liquidity constraints do not exist. They stated that since in reality these conditions do not exist, the shift level of the supply curve is greater than the shift level of the demand curve. In other words, the landlord does not possess information about the renter's future behavior such as whether the renter will wish to renew the contract, whether the renter will demand a compensation fee for surrendering the premises if the landlord rejects the contract renewal, and if so, how much the renter will demand, and whether the renter will bring a lawsuit in order to win. A risk-adverse landlord sets the rent (including key money and security deposit) assuming the worst-case scenario of the renter demanding a high compensation fee for surrendering the premises. On the other hand, there are many cases in which the renter surrenders the rental without demanding a compensation fee even when the landlord rejects contract renewal. For the renter, there are no sources of consumer finance that require no collateral and offer low interest rates, and it is rare for a renter to pay rent for a rental property using consumer finance. They state, therefore, that the degree of leftward shift of the supply curve is larger than that of the demand curve.

6.2.3.3 Expected Counterargument

The following counterargument can, however, be made. Landlords are in a position to know from experience how long the residence period will be from the nature of the tenants (for example, if the tenant is a single person or a family with children) and how much the renewal rent will be restricted based on judicial precedents. Also for renters, in recent years, low-interest-rate consumer finance has been more readily available. If there is a renter protection system and the landlord can receive the future restricted rent portion or compensation fee for the surrender of premises in advance, there should be excess funds compared with if the renter protection system did not exist, and these excess funds should be supplied to renters through the workings of finance. In addition, although uncertainty may exist with respect to rent restrictions or compensation payment for the surrender of premises and the landlord may be risk-averse, the renter himself may wish to take the same level of risk. Taking these factors into consideration, the difference in degree of shifts between the supply curve and the demand curve and the effect of supply restriction owing to the renter protection system should be negligible.

In order to clarify this point, is empirical analysis ultimately necessary?

6.2.3.4 Reasons for the Rise of Equilibrium Rent and Fall of Equilibrium Volume of Rentals

The answer is "not necessarily." The following counter-counterargument can be made.

The amount of renewal rent or compensation payment for the surrender of premises is not immediately determined. Before reaching this point, there is prolonged negotiation between the landlord and the renter, and in some cases, both parties obtain lawyers and the final amount is not set until the court decides. In such cases, both the landlord and the renter incur large transaction costs.

Even if the asymmetry of information and liquidity limitations were small and the difference between the degree of shift of the supply curve ($S_0 \rightarrow S_1$) and the degree of shift of the demand curve ($D_0 \rightarrow D_1$) owing to the existence of the renter protection system were small, in order for the renter to take advantage of and receive the benefit of the protection system and for the landlord to minimize its loss as much as possible, both the renter and the landlord must incur large transaction costs. Consequently, the supply curve will shift further to the left ($S_1 \rightarrow S_2$) and the demand curve will shift to the left ($D_1 \rightarrow D_2$).

Owing to the existence of the renter protection system, the equilibrium rent will definitely rise and the equilibrium volume of rentals will definitely decrease (Figure 6.2).

6.2.4 Legislation Policy Regarding House Lease Law

6.2.4.1 First Best Solution: Elimination of the Just Cause System

The just cause system was not derived from a noble principle of protecting the weak based on social justice. It was merely a supplemental measure to make rent control effective in order to reduce social unrest such as riots in abnormal times during and after the war. For this reason, the person responsible for the legislation should have removed the just cause system when, in 1950, buildings built after July 11, 1950,

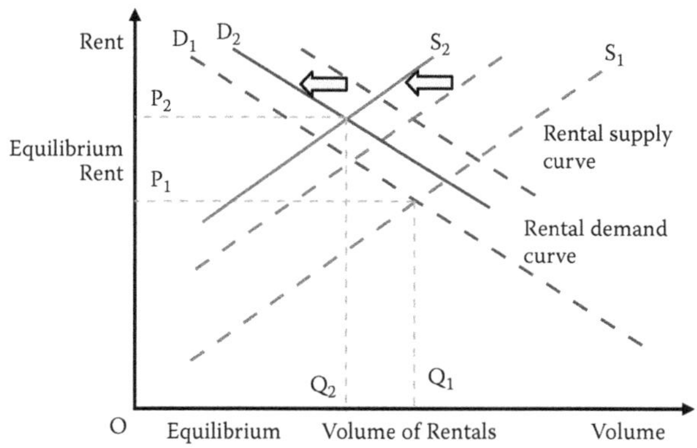

FIGURE 6.2 Effect of implementation of the renter protection system on the market (detailed explanation).

were to be regarded as exempt from the Rent Control Act—in other words, when rent control was eliminated.

The just cause system, which does not allow the rejection of lease renewals without the payment of high compensation fees for the surrender of the premises, and which was established through judicial precedents after the war and enshrined in law under the 1992 law amendment, was a Frankenstein born of judges who were either ignorant of the fact that newly built buildings were exempt from the Rent Control Act or unclear about its legislative intent, and who regulated renewal rent in order to save renters based on benevolence, which was irresponsible, given that the judges would not have been personally affected financially.

6.2.4.2 Second Best Solution: Revision of Renewal Rent in Line with the Market

Even if the person in charge of legislation were to regard rental contracts with limitations on termination as worthwhile and were to continue the just cause system, in disputes regarding rent revisions, the judges should have appropriately managed the rent increase/decrease demand rights so that the renewal rent changed in line with market rent. The courts, however, distortedly interpreted the rent increase/decrease demand rights even for newly constructed buildings exempt from the Rent Control Act, and allowed the regulation on renewal rent increase to continue.

It is true that housing shortages in Japanese cities after the war were serious. Even if a tenant who had been living in the same property for a long period of time protected by the limitations on lease terminations had his lease renewal rejected and was forced to surrender the premises, if the renewal rent had been changed in line with market rent, in other words, if the tenant had the ability to pay market rent, he would still have been able to find a similar place nearby and move there. He would not have ended up in a situation where he could not find housing. If legal precedents had not been established that stated that unless a high compensation fee for the surrendering of the premises was paid, just cause is not constituted, and such legal precedents would not have been written into the law.

As a second-best solution, we should amend the case law so that renewal rent is changed appropriately in line with market rent, and implement a legal rule so that renewal rent can be changed in line with market rent, as in European countries.*

6.3 COMPARATIVE STUDY OF LOWS REGARDING THE JUST CAUSE SYSTEM

Those who defend the just cause system stress that "even in developed countries, a restriction similar to the Japanese just cause system and a similar form of rental contracts comprised of renewal rent restrictions which are there to prevent excessive profits by greedy landlords exist." In this section, we will analyze the reality of these legal systems.

6.3.1 Comparison with Just Cause Systems in European Countries

6.3.1.1 England

6.3.1.1.1 Limitation on Lease Terminations

The types of rental housing contracts that exist in England are (1) guaranteed short-term rentals (assured shorthold tenancy (AST)), whose contract terminates at term expiration (a termination notice is, however, required 2 months prior to term expiration, and is equivalent to Japan's fixed-term lease contract), and (2) guaranteed lease (assured tenancy (AT)), which requires a just cause for the landlord to reject lease renewals. In the case of AT, as long as one of the seventeen limited requirement items listed and written in law is met, just cause is created. The sufficiency of just cause is determined only by the objective requirements of the landlord, and the situation of the renter is not taken into consideration.†

---

* *Market rent* as stated here means rent of similar type of housing in the vicinity, and rent that the landlord would most likely receive if he leased the rental property to someone new. If a certain number of years/months have passed following construction of such rental, the rent level must be lower to incorporate the age and deterioration and should not be the market rent of a newly constructed rental.

In the light of overseas empirical research, however, there is a tendency for renewal rent to become cheaper than market rent. I believe this is because, as mentioned earlier, in order for the landlord to find new renters, he would incur search costs and would need to ask for a risk premium owing to the asymmetry of information. On the other hand, if the renter is good-natured and has been in a contractual relationship from the past, the landlord can reduce the rent by the above amount that he would have incurred otherwise.

† Under the Housing Act Part I and Part II, Schedule 2, as an example, the following are stipulated as being just cause: own use (Ground 1), foreclosure (Ground 2), and alternative housing (Ground 9).

Ground 1: "The property had been used as a dwelling by the lessor before the start of the lease, and the lessor had notified the lessee in advance that the lessor might demand recovery of the premises for its exclusive use in the future, or the lessor requests the property for its own use and the court approves that that is appropriate."

Ground 2: "In the event the property had been provided as collateral prior to the lease contract, the property may rightfully be disposed of under the 1925 Real Estate Law, the collateral owner requests the recovery of exclusive rights in order to dispose of the property, and these notices are provided appropriately. If this had been notified to the lessee prior to the start of the lease, or if the court determines this as appropriate."

Ground 9: "In the event alternative housing is already prepared for the lessee or will be prepared by the time exclusive recovery becomes effective."

6.3.1.1.2 Rent Control

Both the initial rent and renewal rent are subject to the official rent control. In other words, the parties to the rental contract can request the rent inspector to set a fair rent. The rent set by the inspector becomes the registered rent, and receipt of rent that exceeds this amount becomes prohibited.*

6.3.1.1.3 Practice of Paying Compensation Fees, etc.

As a reflective effect of the fact that initial rent is subject to official rent control, payment and receipt of key money are prohibited. In addition, the act of paying compensation fees to make the renter give up his lease rights is considered an obstructive act and a criminal offense, and in judicial proceedings regarding lease terminations, the court cannot use monetary payments as criteria for its decisions.

6.3.1.2 France

6.3.1.2.1 Limitation on Lease Terminations

Although limitations on terminations exist in rental housing contracts in France, renewal rejection causes are listed on a limited basis, and just cause is created only with the objective requirements of the landlord. The situation of the renter is not taken into consideration (Law Regarding the Rights and Responsibilities of Lessors and Lessees, June 22, 1982, Law No. 526, hereafter referred to as the 1982 Law). It should be noted that next to own use (1982 Law, Article 9, Clause 1†), sale to a third party (1982 Law, Article 1, Clause 1‡) is also stipulated as just cause for rejecting renewals.

6.3.1.2.2 Rent Control

In the 1982 Law, the initial rent is left freely up to the parties. There are, however, rules for rent revisions. In other words, the National Committee of Lease Relations decides the increase/decrease percentage that should be applied to rent at the time of rental contract execution or renewal.§

---

* With respect to fair rent, Article 70 of the Rent Law (1977) states that "the rental demand for housing in the area must be assumed to become greater than supply." The system is not stipulated on the assumption that landlords will take advantage of their monopolistic position and take excessive profits, but on the assumption that the market mechanism is working. In calculating the actual fair rent, the rent inspector utilizes a comparative rent method in which he compares the rents set for similar housing. In addition, it is considered normal to incorporate the building, structure, and repairs, etc., in calculating the rent. Please refer to Uchida (1985, p. 130).

† "Termination of a lease contract can be requested if the purpose is to recover such property in order for one to live in it himself, or for his spouse, lineal ascendant, or lineal descendant to live in it, or for his spouse's lineal ascendant or lineal descendant to live in the property."

‡ "The lessor may not renew the lease contract at time of expiration of the original contract or the renewed contract if the sole purpose is to sell the housing."

§ The National Committee of Lease Relations is comprised of representatives of various national organizations that organize lessors and lessees, respectively (Article 35), and discusses and executes the Fair Rent Agreement every year for the purpose of deciding the maximum increase/decrease percentage that should be applied to rent at time of rental contract renewals (Article 51). If the government sets such maximum decrease/increase, this will be applied on a mandatory basis to all rental contracts. Please refer to Inamoto (1985, pp. 16–17).

6.3.1.2.3 *Practice of Paying Compensation Fees, etc.*

Since the 1982 Law stipulates the prohibition of transfer of the right to lease in principle (accepted under special provisions; Article 15), and compensation fees are not a decision factor in determining just cause, the value of such right is not particularly evaluated or demanded. Though there are situations in which renters who accept voluntary surrender make monetary demands, the amount is limited to the addition of a small amount over the purchase price of carpet and wallpaper (Inamoto, 1985, p. 16).

6.3.1.3 Germany

6.3.1.3.1 *Limitation on Lease Terminations*

There are two types of rental housing contracts: (1) a fixed-term contract that has a contract term of 5 years or less and terminates at term expiration, and (2) a contract with limitations on termination. With respect to the latter, renewal rejection causes are listed on a limited basis. Own use (Civil Code, Article 564B, Clause 2.2*) and rebuilding owing to decay (Civil Code, Article 564B, Clause 2.3[†]) are naturally considered just causes.[‡]

6.3.1.3.2 *Rent Control*

For general private rentals, initial rent is not restricted and, in principle, is decided on the basis of a free agreement.[§]

With respect to revision of renewal rent, the Rent Control Law stipulates a comparative rent method. Rent increase is allowed provided the new rent does not exceed the regionally customary comparative rent and the percentage of raise in the past 3 years does not exceed 30%.[¶]

---

* "In the event that the lessor requires such space as living space that will be used for the lessor himself, a household member, or a family member."

† "In the event that economic use of the real estate is significantly obstructed owing to continuation of the lease relationship, and the lessor will likely incur serious losses as a result."

‡ The renter may demand continuation of the lease relationship if the renewal rejection based on just cause is "cruel" for the renter. If an agreement is not reached between the landlord and the renter, the court decides the term and rent. Unless, however, the renter has a special situation, as long as alternative housing is provided, it will not be deemed cruel.

§ Under Economic Criminal Law, Article 5, however, except under special circumstances rent is not allowed to exceed 20% of regionally customary comparative rent and, under Federal Constitutional Court judicial precedents, if the rent exceeds 50% of comparative rent, it will be considered an act of excessive profit under Criminal Code Article 302a.

¶ Regionally customary comparative rent is defined as "a regionally customary price that has been agreed for the past 3 years or one that has been changed, for a lease of housing that has comparative style, size, facilities, quality, and condition in such city/town or in a comparative city/town" (Rent Control Law, Article 2 (1)).

The landlord must request rent increase in writing to the renter and indicate the reason for the increase. One of the following reasons is required: (a) utilize comparative rent in the "standard rent chart" that is applied to social housing, (b) submit an official expert appraisal, or (c) indicate actual rent cases for three properties that are under the same conditions as the lessor's rental. Please refer to Hirowatari (1985, pp. 86–87).

6.3.1.3.3 Practice of Paying Compensation Fees, etc.

There is no tradition of paying key money in Germany either. During the housing shortage period from the end of the war to the 1950s, there was a practice of paying a compensation fee for terminations owing to the transfer of lease rights. This appeared, however, to have disappeared thereafter as the housing situation improved (Hirowatari, 1985, p. 84).

6.3.2 Views on Reevaluating the Just Cause System in Japan

6.3.2.1 Only the Situation of the Landlord Should Be Considered

In England, France, and Germany, there is a type of rental contract with limitations on termination. In either case, a specific list of limited requirements needed to satisfy just cause, which is necessary for landlords to reject lease renewals, is written in law. There is no comparative measurement between the landlord and renter, and only the situation of the landlord is taken into consideration. Typically, in the event that "own use" or "rebuilding" is required, just cause is satisfied without any conditions.

6.3.2.2 Renewal Rent Revision Rule in Line with Market Rent

In England, France, and Germany, at least with respect to revision of renewal rent, there is rent control that requires compliance with the rules set under law. Renewal rent is guaranteed to be revised in line with market price. These revision rules not only protect renters by avoiding price increase demands by greedy landlords, but also are implemented and managed by taking into consideration the fact that it becomes difficult to revise rent in line with market rent since landlords without just cause cannot reject lease renewals and price negotiation power is weakened.

6.3.2.3 No Custom of Paying Compensation Fees

If renewal rent is revised in line with market rent, a lease gain will not be obtained by renters, and renters with ability to pay market rent will have no difficulty finding housing even if they are forced to surrender the premises. Therefore, there will be no need to evaluate the payment of compensation fees for the surrendering of premises as one of the requirements to satisfy just cause, and there is no custom of actually paying such compensation fees. Rather, it would become common for the payment of compensation fees to become a criminal offense.

6.4 EMPIRICAL ANALYSIS OF THE REQUIREMENTS TO SATISFY JUST CAUSE

In this section, we will analyze the requirements needed to satisfy just cause in the courts by analyzing judicial precedent data surrounding just cause for rentals. The judicial precedent data that we analyze are the 176 cases collected from Sawano

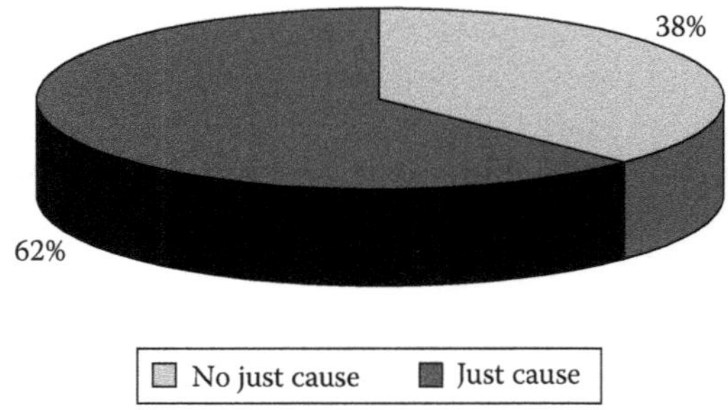

FIGURE 6.3 Status of approvals of just cause.

(2001) and the Commercial Law Center, Inc. (2004).* Out of these 176 cases, just cause was approved for 110 cases, which accounts for 62%. In 66 cases, or 38%, just cause was rejected (Figure 6.3).

6.4.1 Analysis of Requirements That Satisfy Just Cause

6.4.1.1 Overview of Rental Contracts, etc.

At the time of the court ruling, an average of 27.0 years had passed since the time of rental contract execution, and the average latest monthly rent was 378,000 yen (Table 6.1). When comparing those who had just cause with those who had not, the number of years that had passed was approximately the same. In terms of monthly rent, however, those that did not have just cause stood at 444,000 yen, considerably greater than the monthly rent of 338,000 yen of those who did have just cause.

6.4.1.2 Status of Proposal/Payment of Compensation Fees and Satisfaction of Just Cause

The amount proposed by landlords as compensation fee for the surrendering of premises is an average of 16.83 million yen but varies greatly with a maximum amount of 600 million yen. The amount of compensation fee ruled by the court as necessary to approve just cause was an average of 22.61 million yen with a maximum of 800 million yen.

When those with just cause are compared with those without just cause, the average proposal amount for those with just cause is 21.94 million yen, which is greater

* Sawano (2001) collected and organized a total of 137 judicial precedents on just cause for land and house leases from the 1980s that are recorded in the *Collection of Judicial Precedents*, etc., categorized the requirements to satisfy just cause (own use, rebuilding owing to decay, effective use, etc.), and organized the judicial precedent summary, case summary, and court decision summary, etc., for approved cases and dismissed cases. Here, we have analyzed 93 judicial precedents concerning rentals. The judicial precedents were those decided by district courts and high courts nationwide between 1980 and March 2000.

The 83 cases presented in a collection of case precedents by the Commercial Law Center, Inc. (2004) are cases whose dispute point was the existence of just cause in a rental contract, and they were taken from complete records of civil lawsuit cases for which decisions by the courts of first instance had been made between the 3 years from April 2000 to March 2003.

TABLE 6.1
Overview of Rental Contracts

		Combined				No Just Cause				Just Cause			
		Number	Average	Min	Max	Number	Average	Min	Max	Number	Average	Min	Max
Years of residence (years)		165	27.0	2.8	71.6	64	26.5	3.0	68.4	101	27.4	2.8	71.6
Monthly rent (in 10,000 yen)		161	37.79	0.45	1,250.00	61	44.38	1.00	500.96	100	33.77	0.45	1,250.00
Compensation fee	Proposal by landlord (in 10,000 yen)	176	1,683	0	60,000	66	832	0	8,000	110	2,194	0	60,000
	Judgment amount (in 10,000 yen)	176	2,261	0	80,000					110	3,114	0	80,000
	In months of rent (months)	159	144	0	5,128	61	71	0	667	98	189	0	5,128
	Ratio versus net rent paid	49	93%	0%	1,554%	14	0%	0%	551%	35	96%	0%	1,554%

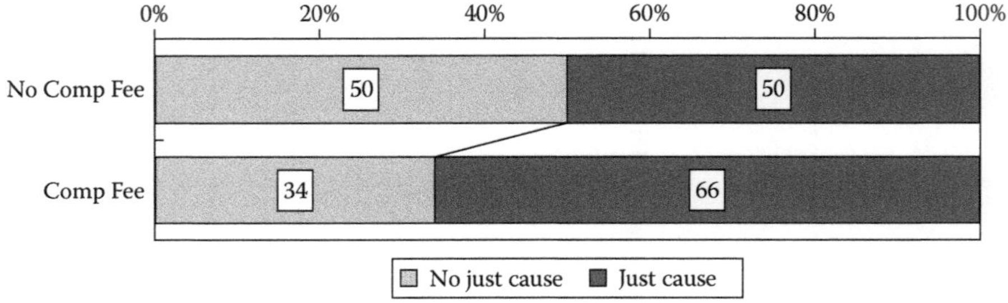

FIGURE 6.4 Status of approvals of just cause based on whether or not compensation fees were paid.

than the average proposal amount of 8.32 million yen for those without just cause. In addition, the average compensation fee ruled by the court is 31.14 million yen, which is 42% greater than the proposal amount. If we divide this court-decided compensation fee by the latest monthly rent and view it from the perspective of number of months, the overall average is 144 months of rent (approximately 12 years), with over 189 months' (approximately 16 years) rent for those with just cause and 71 months' (approximately 6 years) rent for those without just cause. Naturally, the greater the compensation fee, the easier it becomes to satisfy just cause.

If you look at the relationship of total rent payments with compensation fee payment from the 49 cases whose total payments can be estimated from court records, the compensation fee makes up 93% of total payments.

In addition, in cases where the landlord proposed payment of a compensation fee, just cause was satisfied in 66% of cases. If, however, there was no such proposal, just cause was satisfied in only 50% of the cases (Figure 6.4). The payment of compensation fee is a strong factor in obtaining approval for the satisfaction of just cause.

The main factors considered by the court in calculating compensation fees for the surrender of premises were the following: (1) price of lease rights; (2) moving costs, in other words various expenses required by the renter in leasing a new property (moving expense, deposit, key money, broker fees, interior construction expense for the new building, difference in rent, etc.); (3) compensation for sales (decline in sales and operating profit, etc.); and (4) rent (total amount of rent received, rent for a specific period of time). Over 50% of cases took into consideration the price of lease rights as a basis for calculating the compensation fee (Figure 6.5).

6.4.1.3 Advantageous Factors for Landlords in Obtaining Approval for Just Cause

6.4.1.3.1 Need for Own Use

In 91% of cases in which landlords required own use for housing and in 81% of cases in which landlords required own use for business, the ruling was that just cause existed. In judicial proceedings, however, in which need for own use is not required to be approved as a requirement to satisfy just cause, only 52% were ruled as having just cause (Figure 6.6). The necessity for self-use for landlords is one of the important requirements that satisfy just cause.

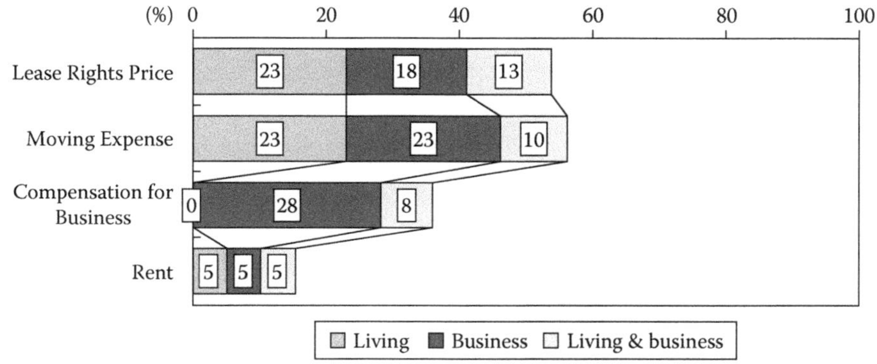

FIGURE 6.5 Factors considered in calculation of compensation fees.

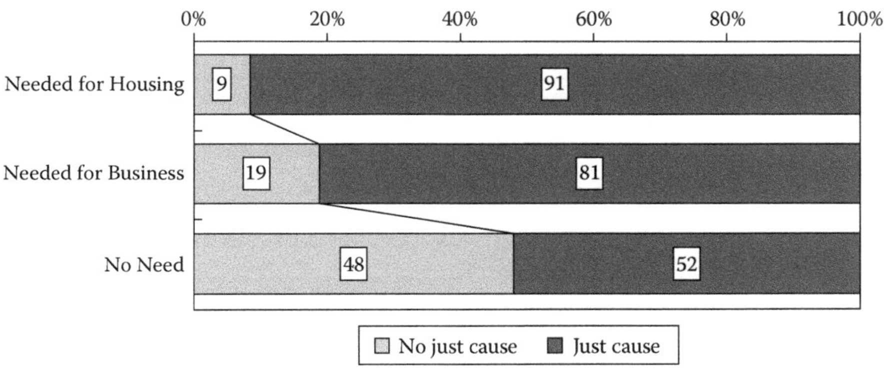

FIGURE 6.6 Status of approvals of just cause based on whether or not the landlord had necessity for own use.

It should, however, be noted that the approval of the landlord's need for own use as a requirement that satisfies just cause is made only after comparative measurement of such factors with the renter's need for own use.

6.4.1.3.2 Need for Rebuilding

In eighteen cases in which it was accepted that rebuilding was necessary due to existing decay and risk of collapse, all cases were ruled as having just cause (Figure 6.7). This is, therefore, a powerful factor in satisfying just cause.

When it was accepted, however, that rebuilding was necessary in order to maintain the effectiveness of the building, only 62% of cases were ruled as having just cause, which is not much greater than the 53% for cases that did not need rebuilding. Consequently, it is not a very important factor in satisfying just cause.

For the 108 cases for which it was accepted that rebuilding was necessary, and in view of the factors that supported this need for rebuilding, for cases in which it was accepted that rebuilding was necessary also from the viewpoint of public interest taking into consideration the regional environment, 92% of them satisfied just cause (Figure 6.8). This is, therefore, a powerful supporting factor in satisfying just cause.

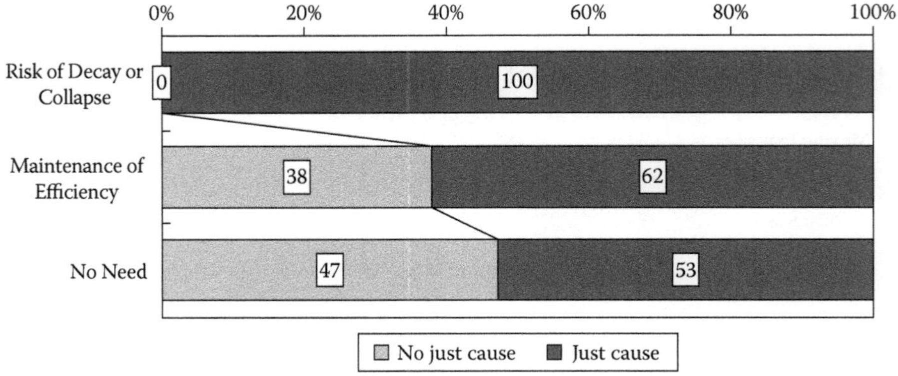

FIGURE 6.7 Status of approvals of just cause based on whether or not rebuilding is needed.

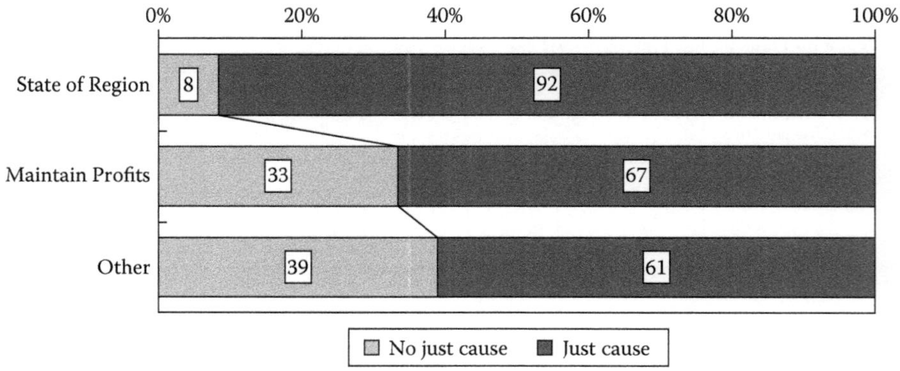

FIGURE 6.8 Status of approvals of just cause based on purpose of rebuilding.

In contrast, if rebuilding is necessary in order to secure profits owing to payment of inheritance tax, etc., or unavoidable business reasons, etc., just cause existed in only 67% of the cases, not much greater than the 61% approved for cases that did not fall into either category. This is not deemed a powerful supporting factor in satisfying just cause.

6.4.1.3.3 *Provision of or Assistance with Alternative Housing*

There were only two cases in which the landlord provided alternative housing to the tenant, and in both cases, it was ruled that just cause existed (Figure 6.9).

On the other hand, when assistance with alternative housing was provided, just cause was ruled to exist in only 53% of the cases, meaning this is not a powerful factor in satisfying just cause.

6.4.1.3.4 *Other Factors*

In the event that it was accepted that from the beginning the rental contract was executed under the agreement that it would be for temporary use or a limited time

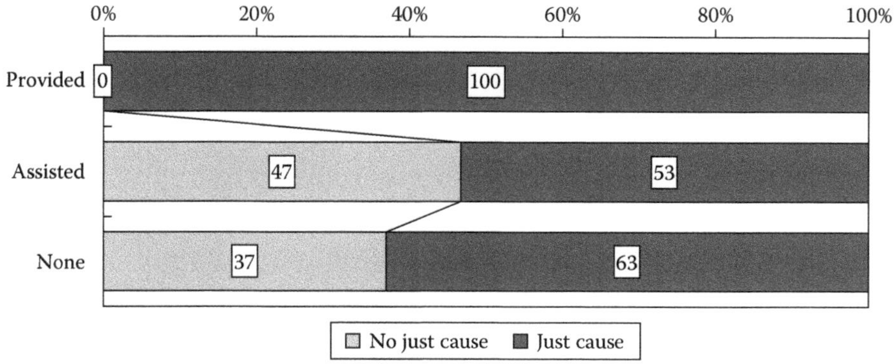

FIGURE 6.9 Status of approvals of just cause based on whether or not alternative housing or assistance with alternative housing was provided.

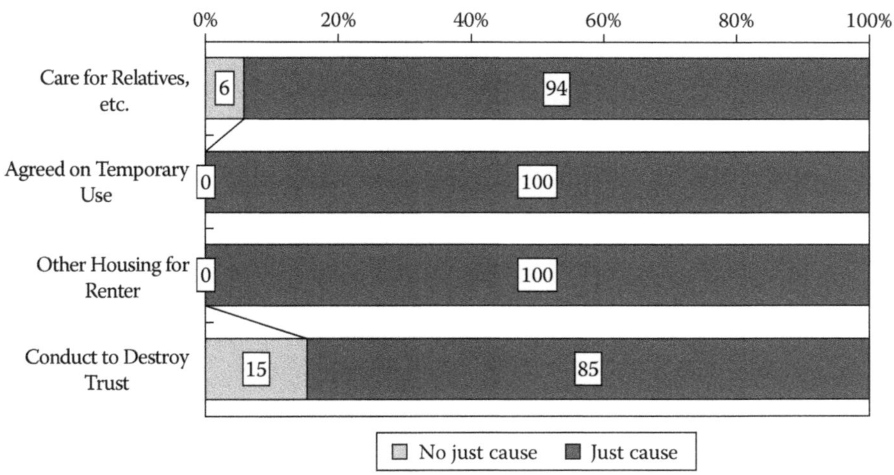

FIGURE 6.10 Status of approvals of just cause based on other factors advantageous to landlords.

period, and there is other housing that can be used by the renter, ultimately all cases were ruled as having just cause.

In cases where it was accepted that own use was necessary since the landlord was to live together with relatives to provide care, etc., to them, and in cases where it was accepted that renters conducted acts that destroyed the relationship of trust, it was ruled in 94% and 85% of cases, respectively, that just cause existed (Figure 6.10). Either is a powerful factor in satisfying just cause.

6.4.1.4 Factors Advantageous to Renters for the Rejection of Just Cause

6.4.1.4.1 Own Use by Renters

If there was a need for own use by renters, just cause was ruled not to exist in 41% of the cases if it was for housing, in 55% of the cases if it was for business, and in 71% of the cases if it was for housing and business.

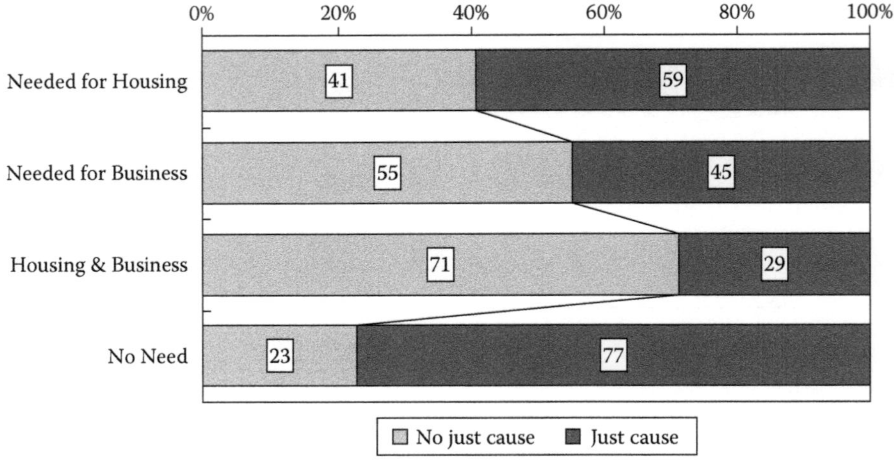

FIGURE 6.11 Status of approvals of just cause based on renter's need for own use.

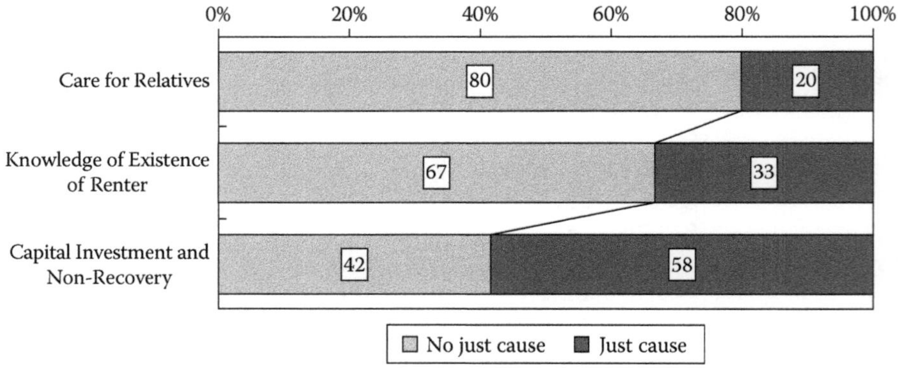

FIGURE 6.12 Status of approvals of just cause based on other factors advantageous to renters.

In contrast, if the need for own use does not exist for the renter, in 77% of the cases just cause was ruled to exist (Figure 6.11). Thus, the need for own use by the renter is a powerful factor in the rejection of just cause.

6.4.1.4.2 Other

Just cause was ruled not to exist in 80% of the cases in which it was accepted that there was a need for own use by the renter for him to live together with relatives to provide care, and in 67% of the cases in which the landlord had purchased the property acknowledging the existence of a renter and had succeeded the lease relationship. These are advantageous factors for the renter in the rejection of just cause (Figure 6.12).

However, when the landlord had invested but not yet recovered capital, only 42% of the cases were ruled as not having just cause, and the remaining 58% were agreed to have satisfied just cause.

6.4.2 Analysis of Compensation Fees to Satisfy Just Cause

Next we analyze the relationship between the just cause satisfaction requirements illustrated in the Land Lease and House Lease Law, Article 28, and the approval or rejection of just cause in judicial precedents, by looking at 176 judicial precedents regarding just cause between 1980 and March 2003.

Specifically, we will call the sample in which just cause was approved in judicial rulings Sample I, and the sample in which just cause was rejected Sample II. Then we will denote the needs of own use by landlords or renters, need for rebuilding or regional environment, background of the lease, and the payment amount of compensation fees for the surrender of the premises as explanatory variables, and estimate the linear function that will determine in which group the sample belongs.

6.4.2.1 Explanatory Variables

With respect to estimating the discriminant function, after the nineteen variables listed in Table 6.3 are entered, the selection or nonselection is decided using the stepwise method. Here, with the exception of LN (years of residence) and LN (compensation fee in number of months), the variables are all dummy variables.

6.4.2.2 Estimate of Linear Discriminant Function

6.4.2.2.1 Estimate Results

The following discriminants were estimated:

Z (discriminant score) =
–0.9356
+1.2661 * Dummy1 (Own use needed by landlord for housing)
+1.0906 * Dummy1 (Own use needed by landlord for business)
+0.8544 * Dummy1 (Own use needed by renter for housing)
+1.0379 * Dummy1 (Own use needed by renter for business)
+1.2846 * Dummy5 (Renter has other available housing)
+1.0884 * Dummy6 (Rental contract with agreement on temporary use)
–0.9262 * Dummy7 (Landlord purchased with knowledge of existence of renter)
+1.7804 * Dummy8 (Rebuilding is needed due to decay and risk of collapse)
+0.5944 * Dummy8 (Rebuilding is needed in order to maintain efficiency of building)
+1.3861 * Dummy10 (Rebuilding has public benefit considering the regional environment)
+0.2177 * Dummy11*LN((Compensation fee payment)/(monthly rent))

6.4.2.2.2 Probability of Successful Prediction

This discriminant predicts the following: (1) if the discriminant score is positive, there will be a ruling that approves just cause, and (2) if the discriminant score is negative, there will be a ruling that rejects just cause.

For the 176 judicial precedents, as shown in Table 6.2, we can see that applying this discriminant will correctly predict the approval or rejection of just cause in 83.3% of the cases (i.e., 147 cases).

TABLE 6.2
Probability of Successful Prediction

Characteristic Value	Canonical Correlation	Group Center of Gravity	
0.924947	0.693185	No just cause	−1.202
		Just cause	0.759

		Estimate Group				
		No Just Cause		Just Cause		Total
Original data	No just cause	52	83.90%	10	16.10%	62
	Just cause	17	17.00%	83	83.00%	100

Note: Prediction success rate = 83.3%.

6.4.2.2.3 View on Precision

The first reason why it is difficult for the discriminant to estimate with 100% success is because there are individual situations that affect the decision of the courts other than the explanatory variables used as factors that affect the ruling of whether or not just cause exists.

The second reason is because here we use dummy variables for factors that affect the ruling of whether or not just cause exists. When, however, courts decide the existence of just cause they consider, for example, not only the mere need for own use, but also the degree. In the explanatory variables, the precision of the estimate formula is reduced by the amount of such information being removed.

The third reason is because even if the two reasons above could be removed, basically in judicial proceedings there are no objective standards in calculating the payment amount of compensation fees, and it should be noted that it differs depending on the judge.

6.4.2.2.4 Analysis Using Discriminant Function

Out of the 100 cases in which it was ruled that just cause was satisfied, these discriminant functions determined that 83 cases satisfied just cause and 17 cases did not satisfy just cause (prediction success of 83%). Looking at the 83 cases in which just cause was satisfied, we see that the actual payment amount of compensation fees decided for the surrender of the premises was on average 201.86 months' rent and 26.03 million yen (Table 6.4).

In contrast, if we calculate the payment amount of compensation fee at the discriminant score of zero, in other words, at the dividing point at which the courts decide whether or not to approve just cause, the amount is only on average 7.43 months' rent and 570,000 yen. In addition, for 28 cases (about one third of the overall cases), it is estimated that even if the compensation fee offered by the landlord were zero yen, the courts would approve the satisfaction of just cause.

For these 83 cases, compared with the amount that was to be the minimum required amount to satisfy just cause, the landlords paid compensation fees that were up to 130 times greater than that amount.

TABLE 6.3
List of Explanatory Variables

	Description	Combined		No Just Cause		Just Cause	
		Average Value	Standard Deviation	Average Value	Standard Deviation	Average Value	Standard Deviation
Dummy Variables	Needed by landlord for own use for living	0.2065	0.4061	0.0500	0.2198	0.3053	0.4630
	Needed by landlord for own use for business	0.1032	0.3052	0.0500	0.2198	0.1368	0.3455
	Needed by landlord for own use for caring for relatives, etc.	0.1097	0.3135	0.0167	0.1291	0.1684	0.3762
	Landlord provided alternative housing	0.0194	0.1382	0.0000	0.0000	0.0316	0.1758
	Landlord assisted in finding alternative housing	0.0839	0.2781	0.1000	0.3025	0.0737	0.2626
	Needed by renter for own use for living	0.1548	0.3629	0.2000	0.4034	0.1263	0.3340
	Needed by renter for own use for business	0.3742	0.4855	0.5667	0.4997	0.2526	0.4368
	Renter has other available housing	0.0452	0.2083	0.0000	0.0000	0.0737	0.2626
	Needed by renter for own use for caring for relatives, etc.	0.0258	0.1591	0.0500	0.2198	0.0105	0.1026
	Conduct by renter that destroyed trust	0.0774	0.2681	0.0333	0.1810	0.1053	0.3085
	Lease contract with agreement of temporary use	0.0839	0.2781	0.0000	0.0000	0.1368	0.3455
	Landlord purchased with knowledge of existence of renter	0.1226	0.3290	0.2000	0.4034	0.0737	0.2626
	Renter invested capital that has not been recovered	0.0581	0.2346	0.0500	0.2198	0.0632	0.2445
	Rebuilding is necessary owing to risk of decay or collapse	0.0903	0.2876	0.0000	0.0000	0.1474	0.3564
	Rebuilding is necessary to maintain building efficiency	0.4968	0.5016	0.5000	0.5042	0.4947	0.5026
	Rebuilding has public benefit owing to regional circumstances	0.1419	0.3501	0.0167	0.1291	0.2211	0.4172
	Rebuilding necessary in order to secure profits	0.0774	0.2681	0.0667	0.2515	0.0842	0.2792
Other	LN (year of residence)	5.5127	0.7755	5.5031	0.7765	5.5188	0.7789
	LN (compensation fee for the surrender of premises (in months of rent))	4.0194	2.4956	3.3106	2.4611	4.4671	2.4245

TABLE 6.4
Required Compensation Fee Based on Discriminant Function and the Actual Compensation Fee Paid

	Required Compensation		Compensation Actually Paid	
Category	Months	Amount (thousand yen)	Months	Amount (thousand yen)
Ruled as just cause (83 cases)	7.43	578	201.86	26,034
Ruled as no just cause (17 cases)	740.56	75,501	104.65	28,649

On the other hand, with respect to the 17 cases that were incorrectly determined as not having just cause based on the discriminant, while in the actual proceedings just cause was deemed satisfied, naturally, although the actual average compensation fee that was paid was 28.65 million yen, the estimated required amount to satisfy just cause was high, at approximately 75.5 million yen.

By using discriminant functions in this way, the minimum compensation fee payment can be calculated for the current courts to decide that just cause is satisfied.

6.4.3 Calculation Method of Compensation Fee Using Discriminant Function

By using a discriminant function and calculating the payment amount of compensation fee for the surrender of the premises so that the discriminant score is exactly zero, we can determine the formula for calculating the compensation fee.

The specific calculation method is as follows.

6.4.3.1 Calculation Method of Compensation Fee

Calculate the discriminant score by using the discriminant function. If the calculated discriminant score is a positive value, it is predicted that the court will rule that just cause exists with zero compensation fee.

If the calculated discriminant score is a negative value, the payment amount of compensation fee that the court will rule as necessary to satisfy just cause can be calculated by use of the following formula:

$$\text{Payment amount of compensation fee} = (\text{Monthly rent})^*\exp(-Z/0.2177 - 1)$$

6.4.3.2 Compensation Fee in the Event That Renter's Situation Is Not Considered

The Japanese just cause system decides the satisfaction of just cause after comparatively measuring the individual situations of the landlord and the renter. But, if they were not to measure against the renter and consider only the situation of the landlord, what would happen to the compensation fee for the surrender of the premises that is ruled by the court as necessary to satisfy just cause? If we were to estimate this by

TABLE 6.5
Compensation Fee Necessary to Satisfy Just Cause (with Own Use or Rebuilding)

Circumstance of Landlord	Circumstance of Renter	Required Compensation (in months)
Own use needed for living	Not considered	0.00
Own use needed for business		0.00
Leased based on agreement of temporary use		0.00
Rebuilding needed owing to risk of collapse		0.00
Rebuilding needed owing to risk of collapse (Public benefit based on regional environment)		0.00
Rebuilding needed to maintain efficiency		1.76
Rebuilding needed to maintain efficiency (Public benefit based on regional environment)		0.00

using the compensation fee estimate method that uses the discriminant function, it would be according to that shown in Table 6.5. Here, when own use by landlord or rebuilding is necessary, the only time the discriminant score becomes negative is when, despite the fact that rebuilding is necessary in order to maintain the efficiency of the building, it is not determined that rebuilding is necessary also from the perspective of public interest considering the regional environment. Otherwise, all other discriminant scores are positive values, and estimate that just cause will be accepted as satisfied without the payment of compensation fees.

Even if the reason for rebuilding is just for the maintenance of efficiency, it is estimated that satisfaction of just cause will be accepted by paying 1.76 months' rent as compensation fee.

In other words, even on the basis of the decision standards of the Japanese courts, if we were not to take into consideration the individual circumstances of the renter in the event that own use or rebuilding is necessary, it is estimated that just cause would be satisfied with almost zero payment of compensation fees for the surrender of the premises.

Compensation fees for the surrender of the premises should not be required for own use or rebuilding, and that in itself should be a sufficient requirement for just cause.

6.4.3.3 Compensation Fees in the Event the Reason Does Not Fall under Own Use or Rebuilding Requirement, etc.

If the reason does not fall under the own use or rebuilding requirement, how much compensation fee should be required to allow the satisfaction of just cause? Should the amount of required compensation fee be equal to the price of lease rights?

Table 6.6 provides an estimate of compensation fee payments based on whether or not the renter requires the property for his own use or not, in the event that the landlord requires the property for his own use, estimated using the discriminant function based on the decision standards of current courts.

TABLE 6.6
Compensation Fee Necessary to Satisfy Just Cause (without Own Use or Rebuilding)

Circumstance of Landlord	Circumstance of Renter	Required Compensation (in months)
Own use needed for living	Not considered	0.00
Own use needed for living	Own use needed for living	4.08
Own use needed for business	Not considered	0.00
Own use needed for business	Own use needed for business	9.48

It is estimated that if the landlord requires the property for his own use for living, (1) if the situation of the renter is not considered, the court will decide that just cause is satisfied with zero compensation fee; however, (2) if the fact that the renter needs the property for his own use for living is taken into consideration, the court will decide that just cause is satisfied only with the payment of 4.08 months' rent of compensation fee. If the renter needs to dwell in such property, and yet the contract terminates and he is forced to incur moving expenses, the renter will experience a loss by such amount.

Likewise, it is estimated that if the landlord requires the property for his own use for business, (1) if the situation of the renter is not considered, the court will decide that just cause is satisfied with zero compensation fee; however, (2) if the fact that the renter needs the property for his own use for business is taken into consideration, the court will decide that just cause is satisfied only with the payment of 9.48 months' rent of compensation fee. If the renter needs to conduct business in such property, and yet the contract terminates and he is forced to incur moving expenses, the renter will experience a loss by such amount.

Therefore, if the landlord does not have a need for own use or rebuilding, the system would be such that in order to satisfy just cause to reject lease renewals, he would have to pay a compensation fee equal to 4 months' rent in the event that the rental was used for living and 9.4 months' rent in the event that the rental was used for business as payment for moving expenses in order to satisfy just cause.

6.5 SUGGESTIONS FOR LAW REFORM

I hereby summarize my suggestions for law reform based on the above analyses:

1. First best solution: Elimination of the just cause system. Even if the current housing policy designed to protect the weak, such as public housing, is insufficient, completely eliminating the just cause system is simply the first best solution as long as we do not retroactively apply it to existing contracts.* Without the just cause system, renewal rent would not have become

* Existing contracts would need to be handled separately through housing policies, such as prioritizing them in moving into public housing.

so distant from market rent and people would not have experienced difficulty finding new housing.

2. Second best solution: Revision of renewal rent in line with the market. The second best solution is to create the following systems, which are similar to the legal systems existing in European countries, even if we are to continue the just cause system on the understanding that rental contracts with limitations on terminations are meaningful:
   - Specifying and objectifying the requirements for just cause: The requirements that satisfy just cause should be listed on a limited basis and specified in law. In addition, whether just cause exists or not should be decided not through comparing and measuring the landlord and renter, but through taking into consideration only the circumstances of the landlord. In particular, in the event that "own use" or "rebuilding" is required, the system should be such that just cause is satisfied without any conditions.
   - Renewal rent revision rule that is in line with market rent: The price negotiation power of landlords without just cause is weakened since they cannot reject lease renewals, resulting in difficulty in revising rent in line with the market. A mechanism through which renewal rent will be guaranteed to be revised in line with market price should be implemented.
   - Monetary payment standards to satisfy just cause: We should have a system in which just cause will be satisfied by payment of 4 months' rent if the rental property was used for living and 9 months' rent if it was used for business, even if requirements such as "own use" or "rebuilding" do not exist for the landlord.

REFERENCES

Commercial Law Center, Inc. 2004. *Investigation and research report on court cases regarding just cause under rental contracts*. Tokyo.

Fukushima, T. 1998. Is there justification in criticizing fixed term lease rights? In *Fixed term lease rights*, ed. Yasutaka Abe, Yoshihiro Nomura, Hideo Fukui, 200–21. Tokyo: Shinzansha.

Furuyama, H. 1951. *House lease law*. Tokyo: Toyoshokan.

Hirose, T. 1950. *Land lease and house lease law*. Tokyo: Nippon Hyoron-sha.

Hirowatari, S. 1985. *Land lease and house lease system of Germany*. Research report on the land lease and house lease systems in Europe, Housing Research and Advancement Foundation of Japan, 69–120.

Inamoto, Y. 1985. *Land lease and house lease system in France*. Research report on the land lease and house lease systems in Europe, Housing Research and Advancement Foundation of Japan, 11–68.

Iwata, K. 1976. Economic analysis of land lease and house lease law. *Quarterly Modern Economics Magazine* 24:122–38.

Iwata, K. 1994. Economic approaches to urban housing. *Urban Housing Sciences* 8:48–59.

Kanemoto, Y. 1992. Economic analysis of the new land lease and house lease law. *Jurist* 1006:28–34.

Kume, Y. 2003. Issues to be considered under the rental housing law system. *Urban Housing Sciences* 42:55–60.

Sawano, Y. 2001. *Just cause under land and house lease and compensation fees for the surrendering of premises—Collection of judicial precedents/cases*. Tokyo: Shinnippon Hoki.

Suzuki, R. 1959. *Theory on housing rights*. Tokyo: Yuhikaku.

Uchida, K. 1985. *Land lease and house lease system of United Kingdom*. Research report on the land lease and house lease systems in Europe, Housing Research and Advancement Foundation of Japan, 117–16.

Watanabe, Y. 1951. About the landlord's request rights. *Law News* 22:22–33.

7 Qualitative Analysis of Two-Dimensional Urban Employee Distributions in Japan

A Comparative Study with Urban Population Distributions by Means of Graph Theoretic Surface Analysis

Satoru Masuda

7.1 INTRODUCTION

Since the 1950s, extensive researches have been conducted to study urban population distribution. Okabe (1981) has proposed a graph theoretic method in order to make intercity comparison easier, holding the two-dimensional characteristics of the distribution. Using this method, unlike some one-dimensional descriptions of population density including the pioneering work of Clark (1951), the two-dimensional population distributions of the 245 cities in Japan have been studied by Okabe and Masuda (1984). In relation to the size of the population, the shape of the city area, and the subdivision of city area by rivers, some empirical results were reported concerning the emergence of polycentric patterns and doughnut phenomena.

Okabe and Masuda (1984) dealt with one aspect of the distribution of urban inhabitants, the de jure (nighttime) population aggregated by their residential locations. The other aspect is the working (daytime) population, especially the numbers of employees, aggregated by workplaces. Both populations should be examined together to obtain the fundamental knowledge needed in the process of land use and infrastructure planning for housing, commuting and teleworking, development or relocation of urban functions, and so on. From this point of view, this article attempts to make a comparative study of qualitative characteristics of two-dimensional urban employee distributions of the 65 mid-size cities to clarify how and why employee and population distributions differ (Masuda, 1987).

7.2 METHOD OF ANALYSIS AND THE CHARACTERISTICS OF URBAN POPULATION DISTRIBUTIONS IN JAPAN

7.2.1 Graph Theoretic Method

In order to determine the significant global structure of a density surface, the method follows a procedure consisting of the following four steps[1]: (1) estimation of density surface; (2) identification of singular points of the local surface (peaks, bottoms, and cols); (3) graphical expression of topological arrangement of peaks, bottoms, and cols; and (4) extraction of the significant structure by neglecting minor peaks and bottoms (Okabe, 1981).

If we make a population distribution surface by plotting density as a height in a city area on the Earth's surface, three kinds of singular points may exist: peaks are the points that have relatively higher population density than the surroundings; bottoms are the dimple points that have relatively lower population density; and cols are saddle-backed points. In an actual analysis of population distribution, $n_p + n_b - n_c = 1$, where n_p, n_b, and n_c are the number of peaks, bottoms, and cols, respectively. Hence, the proceeding empirical analysis will use the number $n_p + n_b$ of peaks and bottoms as the index of global complexity of the distribution surface.

These singular points are aligned on the ridge lines of steepest ascent going through cols, so that the graph with nodes and links, which correspond to peaks and ridge lines of steepest ascent, respectively, represents the global structure of urban population distribution (hereinafter called population distribution graph G_{POP}). Finally, the significant global structure is obtained by neglecting the peaks/bottoms whose relative heights/depths are less than a significant level Δ. In an empirical examination of population distribution (and proceeding employee distribution), we set $\Delta = 1{,}500/\text{km}^2$, at which level many cities show robust structures.

In Okabe and Masuda (1984), the population data were given by the so-called 1 km × 1 km "mesh data" (standard grid square system used in Japanese spatial statistics) (Administrative Management Agency of Japan, 1973) compiled from the population census held in 1975 by the Japanese Bureau of Statistics (1977). A city was defined as having a connected area consisting of more than five such 1 km × 1 km meshes (quasi quadrats) having more than 2,500 inhabitants. To avoid complication, although the conditions stated above are satisfied, the Tokyo, Osaka, and Nagoya metropolitan regions were excluded from the analysis, because these regions are very wide conurbations and contain several independent core cities. In total, data for 245 cities were obtained, from among which Figure 7.1 shows the population distribution graphs G_{POP} of the 65 cities having more than 100,000 total inhabitants (selectively quoted from Okabe and Masuda (1984) for the sake of comparison with Figure 7.2).

7.2.2 Structural Comparison of Population Distributions by Isomorphism of the Graph

Once the distributions of cities $i = 1, 2, \ldots, m$ are represented by graphs $\{G_i\} = \{G_1, G_2, \ldots, G_m\}$, the global structures can be easily compared. If graph G_i is isomorphic to G_j, we may say that the global structure of the population distribution of city i is qualitatively similar to that of city j. The isomorphism of graphs is initially characterized by the number of nodes, the degree of each peak, and so on. For example, both Okayama and Hakodate in Figure 7.1 are the same in terms of the number of nodes (four) but heteromorphic in structure: the former shows a radial pattern of peak arrangement, and the latter shows a linear one. In addition, the most fundamental dichotomy of $\{G_i\}$ is whether G_i is a tree graph ($n_b = 0$) or a circuit graph ($n_b > 0$), with or without elementary circuits.[2]

Okabe and Masuda (1984) noticed that the most frequently observed tree graphs were one-node, that is, monocentric, cities (Yakagata, Morioka, and the others in the top of Figure 7.1a). Such cities comprised two thirds of the total of 245 cities, but there were only 12 monocentric cities among the 65 cities with population sizes larger than 100,000 (these 65 cities are the objects of the study in Section 7.3).

The increase in the number n_p of peaks indicates the transition from monocentric to polycentric structure. There were no tree graph cities with more than seven peaks. This shows that the population distribution surface with more than seven peaks does not remain simple in the sense that the graph does not remain a tree graph but becomes a circuit graph. Moreover, the existence of the node of degree 3 or more (degree of node is the number of links connecting to the node) may correspond to the radial or so-called "finger" pattern alongside the main roads or railways, but radial patterns without circuits were very rare.

7.2.3 Factors Relating to the Structure of Population Distribution

One of the simple and observable factors that related to the structural change of the qualitative population distribution of a city might be its population size P (see Table 7.1b). In fact, there were distinct changes around P = 100,000 and 200,000. When the population was below the former level, almost all the cities were tree graph. In contrast, when the population exceeded the latter level, the circuit graph became dominant.

Besides population size, the shape of a city might affect the qualitative structure of the population distribution as well. To examine this factor, we used *shape index* $S = 4\sqrt{A}/L$, where A is the area and L is the perimeter of a city.[3] The range of this index is $0 \le S \le 1$, and the compactness increases as S increases. The upper bound and the average of the shape index decrease monotonically with respect to the number of

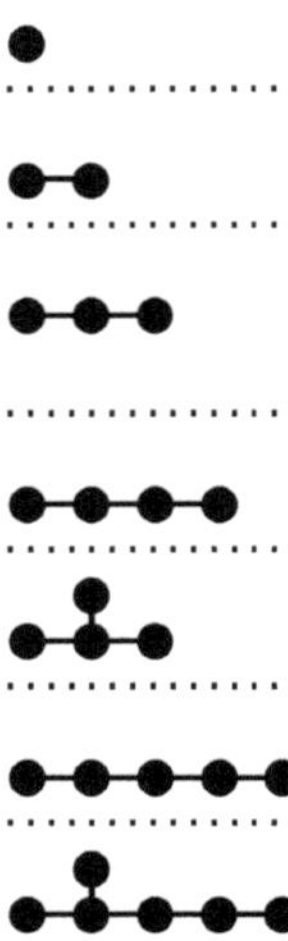

Number of peaks	Name	Number of cities
$n_p = 1$	Yamagata, Morioka, Aomori, Kiryu, Takasaki, Toyama, Fukui, Nagano, Takamatsu, Kochi, Kurume, Oita	12
$n_p = 2$	Koriyama, Hachinohe, Akita, Hitachi, Toyohashi, Hamamatsu, Kanazawa, Himeji, Kure, Beppu	12
$n_p = 3$	Muroran, Otaru, Kushiro, Utsunomiya, Mito, Kofu, Fuji, Okazaki, Otsu, Kashihara, Fukuyama, Miyazaki, Sasebo, Omuta, Kagoshima	15
$n_p = 4$	Hakodate, Maebashi, Numazu + Mishima, Nara, Tokushima	5
$n_p = 4$	Okayama, Takasago, Shimonoseki	3
$n_p = 5$	Naha	1
$n_p = 7$	Niigata	1

(a) The tree graph cities

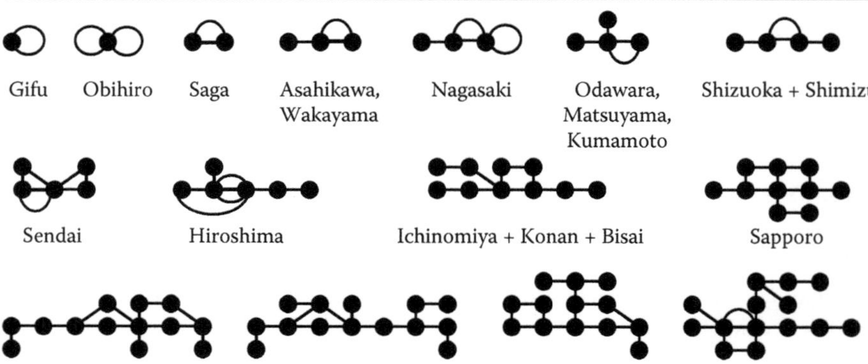

(b) The circuit graph cities

FIGURE 7.1 The population distribution graphs G_{POP} of the 65 cities. The 65 cities having more than 100,000 total inhabitants are selectively quoted from Okabe and Masuda (1984).

peaks. The decline in these values shows that the number of peaks tend to increase as the shape of a city becomes less compact.

In many cities, the area of a city is divided by physical elements such as rivers, hills, and cliffs, and these elements might be related to the topological structure of population distribution. Let a *separation number* be the number of subareas separated by rivers, denoted by N. If the separation number N of a city was more than 5, the doughnut phenomenon was likely to appear, and on average the number of the doughnut phenomena in a city increased as the separation number N increased.

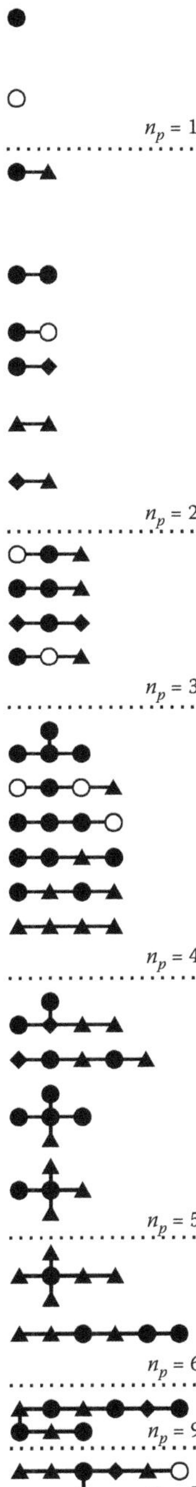

Taxonomic city group:[a]	I	II	III	IV	Total	Name
						Number of cities
	6	–	–	–	19	(Akita, Aomori, Nara, Kochi, Beppu, Miyazaki)
	–	13	–	–		(Otaru, Yamagata, Koriyama, Maebashi, Kofu, Toyohashi, Gifu, Fukui, Himeji, Tokushima, Takamatsu, Saga, Kurume)
	–	1	–	–	1	(Kanazawa)
$n_p = 1$					20	
	2	–	–	–	12	(Matsuyama, Kagoshima)
	–	6	–	–		(Takasaki, Utsunomiya, Nagano, Toyama, Okayama, Nagasaki)
	–	–	2	–		(Odawara, Otsu)
	–	–	–	2		(Kiryu, Takasago)
	2	–	–	–	5	(Morioka, Sendai)
	–	3	–	–		(Niigata, Fukuyama, Sasebo)
	–	3	–	–	3	(Asahikawa, Obihiro, Mito)
	–	1	–	–	3	(Hachinohe)
	–	–	2	–		(Numazu + Mishima, Okazaki)
	–	1	–	–	2	(Omuta)
	–	–	1	–		(Fuji)
	–	–	1	–	1	(Kure)
$n_p = 2$					26	
	1	–	–	–	1	(Sapporo)
	–	2	–	–	2	(Hakodate, Kushiro)
	–	–	1	–	1	(Kashihara)
	–	–	1	–	1	(Wakayama)
$n_p = 3$					5	
	1	–	–	–	1	(Fukuoka)
	1	–	–	–	1	(Kumamoto)
	1	–	–	–	1	(Naha)
	–	1	–	–	1	(Oita)
	–	–	1	–	1	(Muroran)
	–	–	–	1	1	(Hitachi)
$n_p = 4$					6	
	–	1	–	–	1	(Chiba)
	–	–	1	–	1	(Shizuoka + Shimizu)
	–	–	1	–	1	(Shimonoseki)
	–	–	–	1	1	(Hamamatsu)
$n_p = 5$					4	
	–	1	–	–	1	(Hiroshima)
	–	–	–	1	1	(Ichinomiya + Konan + Bisai)
$n_p = 6$					2	
$n_p = 9$	–	–	1	–	1	(Kitakyushu)
$n_p = 7$	–	–	1	–	1	(Kyoto)

The type of peaks[b]

- ● : The tertiary industrial (Type Trd)
- ◆ : The mixed industrial (Type Mix)
- ○ : The government service (Type Gov)
- ▲ : The secondary industrial (Type Snd)

FIGURE 7.2 The employee distribution graphs G_{EMP} of the 65 cities. (a) See Table 7.3 concerning the definition of taxonomic city groups I, II, III, and IV. (b) See Table 7.2 concerning the definition of the types of peaks represented by different symbols.

7.3 THE CHARACTERISTICS OF URBAN EMPLOYEE DISTRIBUTIONS IN JAPAN

In this section, the method introduced above[1,4,2] and the mesh data compiled from the establishment census held in the same year as the population census (1975) are used. Taking the following points into account, we obtain the 65 graphs of employee distribution surface representing the density of the total number of persons engaged in the business establishment.[4]

The first point concerns the definition of a city. In the case of employee distribution, the relocation of factories and wholesalers from the inner city to suburban districts, and the new development of industry parks and merchandise marts in less populated areas, are often observed in some cities. To include these locations in the study, we set up a wider study area: a city defined in Section 7.2.2 for population analysis and its contiguous surrounding meshes (i.e., sharing not only corners but also sides). It is, however, reasonable that a distant site selected for an industrial location should be regarded as an independent peak of employment when it is more than several kilometers away from an existing city.

Second, considering the three points listed below, we select mid- or large-sized cities whose residential population is more than 100,000: (1) the size of a 1 km × 1 km mesh (standard grid square) is rather wide relative to the area of a small-sized city; (2) the number of total employees in a city is smaller than that of the population (see *E*/*P* in Table 7.1); and (3) a more concentrated distribution of workplace population than that of residential population may be supposed.

Putting the difference in plotting symbols of node aside for the present, the 65 graphs represent the 12 kinds of global structures of employee distribution as listed in Figure 7.2 (hereinafter called employee distribution graph G_{EMP}). If the decline of a central business district (CBD) in a city were apparent, like the inner city problem, a G_{EMP} belonged to a circuit graph. In the case of population graph G_{POP}, the doughnut phenomenon appears when the population is greater than 90,000 and becomes dominant when the population exceeds 200,000. In contrast, Figure 7.2 shows that there is no circuit graph G_{EMP} except Kyoto, assuming that a city has more than 200,000 employees.

7.3.1 STRUCTURAL COMPARISON OF EMPLOYEE DISTRIBUTIONS BY THE NUMBER OF PEAKS

Table 7.1a and b shows the numbers of graphs G_{EMP} and G_{POP}, respectively, and the average values of indices of such cities sorted as isomorphic: tree graph cities with one peak, two peaks, three peaks, four peaks, and five or more peaks, and circuit graph cities. We notice from Table 7.1a that the most frequently observed employee distribution graph G_{EMP} is a two-node tree graph, that is, bipolar cities, and the second most frequent is one-node. Specifically, the ratios of two-node and one-node tree graphs G_{EMP} to the total, 65, are about 40% and 30%, respectively. (In contrast, those of population distribution graphs G_{POP} are about 15% and 20%.) At the same time, the average numbers of peaks in G_{EMP} and G_{POP} among the 65 cities are 2.43 and 3.65, respectively. These values show that the more concentrated distribution

TABLE 7.1
Average Characteristics of G_{EMP} and G_{POP} by the Number of Peaks

	Number of Cities	Population Size *P* (Inhabitants)	Employment Size *E* (Person Engaged)	*E*/*P* (%)	Shape Index *S*	Separation Number *N*
a. Employee Distribution Graphs G_{EMP} of the 65 Cities						
Tree graph						
$n_p = 1$	20	159,312	103,167	65.2	0.456	2.35
$n_p = 2$	26	189,064	117,288	63.9	0.368	3.12
$n_p = 3$	5	360,326	195,903	52.2	0.325	3.60
$n_p = 4$	6	368,399	205,429	58.1	0.304	3.33
$n_p \geq 5$	7	485,058	278,355	58.5	0.237	4.29
Circuit graph	1	1,667,780	748,617	44.9	0.204	8.00
b. Population Distribution Graphs G_{POP} of the 65 Cities						
Tree graph						
$n_p = 1$	12	144,358	104,824	73.6	0.465	2.42
$n_p = 2$	10	169,480	112,131	65.1	0.404	2.60
$n_p = 3$	15	159,091	94,749	60.6	0.355	2.80
$n_p = 4$	8	186,923	106,285	56.6	0.324	3.38
$n_p \geq 5$	2	333,521	181,662	56.0	0.302	3.50
Circuit graph	18	511,167	278,159	56.7	0.327	4.06

pattern, the duo- or monocentric structure, is more likely to occur in the distribution of employees than in that of population on average.

In the case of tree graph cities, the highest degree of node in G_{EMP} is 4 (that of G_{POP} is 3), which is to say, Shimonoseki, Hamamatsu, and Hiroshima with nodes of the highest degree show the four-way radial pattern of employee distribution. Figure 7.3 is the cross-tabulation of the numbers of cities by isomorphic types of G_{EMP} and G_{POP}. The constellation shifted to the upper right shows the more concentrated pattern of employee distribution as well.

There is only one circuit graph city in the case of G_{EMP}, and the number n_p of peaks can, therefore, represent most of the global complexity of the distribution surface of urban employees.

7.3.2 Types of Peak According to the Industrial Composition

To examine the workplace characteristics around a peak point, we simply use the numbers of persons engaged (the number of employees) within the mesh (standard grid square) including it. Based on the 10 major industry groups,[5] all 158 peaks of 65 employee distribution graphs G_{EMP} are classified into four *types of peak*, Trd, Mix, Gov, and Snd, according to the rules shown in Table 7.2: Trd, the tertiary industrial type, indicated by a filled circle; Mix, the mixed industrial type, filled diamond; Gov, government service type, open circle; and Snd, the secondary industrial type, filled triangle.

		The Population Distribution Graphs G_{POP}						
		Tree Graphs					Circuit Graphs	Total
	n_p	1	2	3	4	≥5		
The Employee Distribution Graphs G_{EMP} — Tree Graphs	1	*** ***	*** ***	***	***		**	20
	2	*** ***	**	*** *** ***	***	*	*** ***	26
	3			**	*		**	5
	4	*[a]	*[b]	*[c]		*	**	6
	≥5		*[d]		*[e]		*** **	7
Circuit Graphs							*	1
Total		12	10	15	8	2	18	65

FIGURE 7.3 The relationship between the employee and population distribution graphs. Asterisks represent the 65 cities. The following cities have more peaks in G_{EMP} with $n_p \geq 4$ than in G_{POP}: (a) Oita, (b) Hitachi, (c) Muroran, (d) Hamamatsu, and (e) Shimonoseki.

TABLE 7.2
Types of Peak in Employee Distribution Graphs G_{EMP} according to the Industrial Composition[4,5]

Types of Peak	Largest Ratio		Second Largest Ratio
●: Tertiary industrial (type Trd)	One of the tertiary industry groups except government service (4 to 9)	and	One of the tertiary industry groups (4 to 10)[b]
◆: Mixed industrial (type Mix)	One of the tertiary industry groups except government service (4 to 9)	and	One of the secondary industry groups (1 to 3)
○: Government service (type Gov)	Government service (10)		
▲: Secondary industrial (type Snd)	One of the secondary industry groups (1 to 3)[a]		

[a] The major secondary industry groups: (1) mining, (2) construction, (3) manufacturing.

[b] The major tertiary industry groups: (4) wholesale and retail trade, (5) finance and insurance, (6) real estate, (7) transport and communications, (8) electricity, gas, heat supply, and water, (9) services, (10) government service.

Taking the difference between these types of peaks into consideration as well, the original 12 isomorphic groups of G_{EMP} are segmented into 26 subgroups, most of which consist of only one city or precious few cities.[6] (This variety might be related to the citywide industrial composition of employment as described below, in Section 7.3.4.)

Figure 7.2 shows that almost all the cities have at least one peak of type Trd, where the agglomeration of commerce or service industry is significant. Among the cities with two or more peaks, there is a common configuration where a city has one peak of type Trd at CBD and other peak(s) of subcenter in its fringe area. Especially in the case of the 26 two-peak cities, the combination of two peaks where one is type Trd and the other is type Snd represents close to half (twelve) of the cities, which may link to the spatially and functionally differentiated bipolar locations. In contrast, there are only three cities, Hitachi, Fuji, and Omuta, which are known as manufacturing or mining towns, whose employee distribution graphs G_{EMP} consist only of peaks of type Snd (Kure is another discriminating two-peak city whose G_{EMP} is a combination of types Mix and Snd).

7.3.3 Factors Relating to the Structure of Employee Distribution

To examine the relation between such factors as city size and the qualitative structure of employee distribution, Table 7.1 also shows the averages of the following indicators by the number of peaks: population size P, employee size E, ratio of E to P, shape index S, and separation number N. (Table 7.1b shows the values of population distribution graph G_{POP} sorted not by the number $n_p + n_b$ of peaks and bottoms, but by the number n_p of peaks.) The logarithms of population size P and employee size E are depicted in Figures 7.4 and 7.6, respectively, with the number $n_p + n_b$ of peaks and bottoms of G_{EMP} in a city. In these figures, a circle represents the only circuit graph city, Kyoto, and cross-marks represent the remaining tree graph cities.

7.3.3.1 Population Size

Figure 7.4 shows that there is a not-so-strong positive linear relationship with significant changes around $P = 200{,}000$ and 400,000. As the number n_p of peaks in a graph G_{EMP} increases, distinctive gaps in the average population size are observed in between $n_p = 2$ and 3 and $n_p = 4$ and 5. To be precise, the average population sizes of the one-peak, two-peak, three-peak, four-peak, and five-or-more-peak cities are 159,312, 189,064, 360,326, 368,399, and 485,058, respectively (see Table 7.1a).

Moreover, the upper limits to the population sizes of one-peak and two-peak cities are about $P = 300{,}000$ and 550,000, respectively. These limits show that a city cannot maintain a monocentric structure above the former population size, and that a city having more inhabitants than the latter size has at least two subcenters of employee distribution (trilateral structure).

In addition, as the population size of a city increases and exceeds 300,000, the number n_p of peaks of G_{POP} is almost always greater than that of G_{EMP}, and the difference between them tends to increase. This means that the distribution of both urban population and employment become more diffuse as the population size of a city increases, but this trend is less significant in employee distribution. If we consider the number $n_p + n_b$ of peaks and bottoms as an index of surface complexity (i.e., convexo-concave structure), the difference of the complexity between G_{POP} and G_{EMP} becomes more remarkable (Figure 7.5).

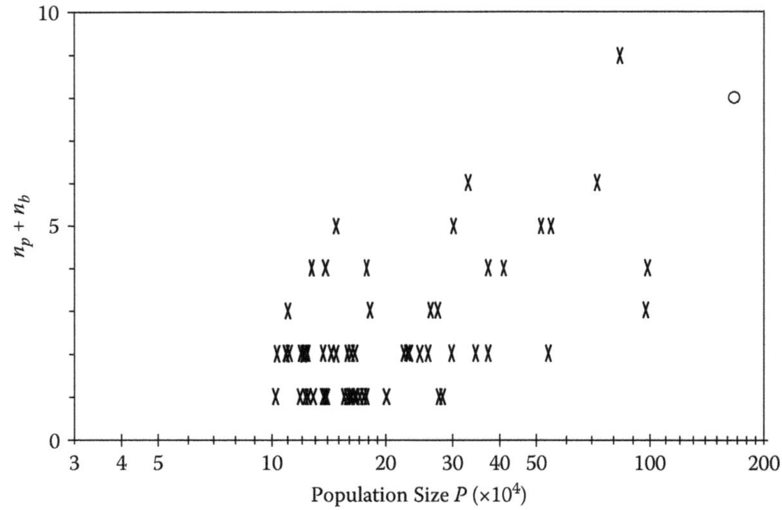

FIGURE 7.4 The relationship between population size P and the number $n_p + n_b$ of peaks and bottoms of employee distribution graph G_{EMP}. A circle represents the only circuit graph city, Kyoto, and cross-marks represent the remaining tree graph cities.

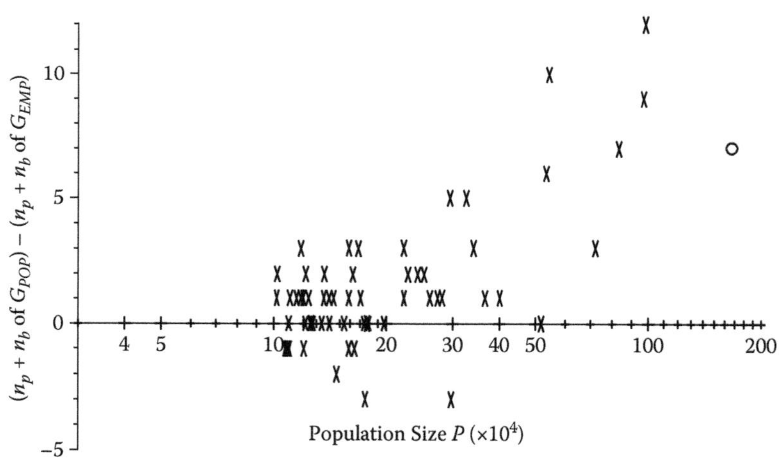

FIGURE 7.5 The relationship between population size P and the difference in the number $n_p + n_b$ of peaks and bottoms between G_{POP} and G_{EMP}.

7.3.3.2 Employment Size

The average ratio of total employees to population is around 60% (see *E*/*P* in Table 7.1). Therefore, compared with Figure 7.4, Figure 7.6 shows a shifted scatter pattern to the left (smaller-size side) by just that much, in which there is a significant change around E = 200,000. As the number n_p of peaks in a tree graph increases, the average employment size increases. The upper limits to the employment sizes of

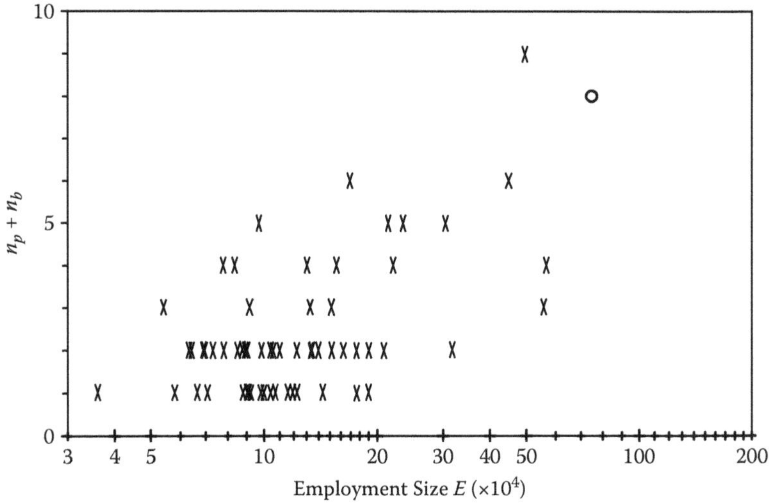

FIGURE 7.6 The relationship between employment size E and the number $n_p + n_b$ of peaks and bottoms of the employee distribution graph G_{EMP}.

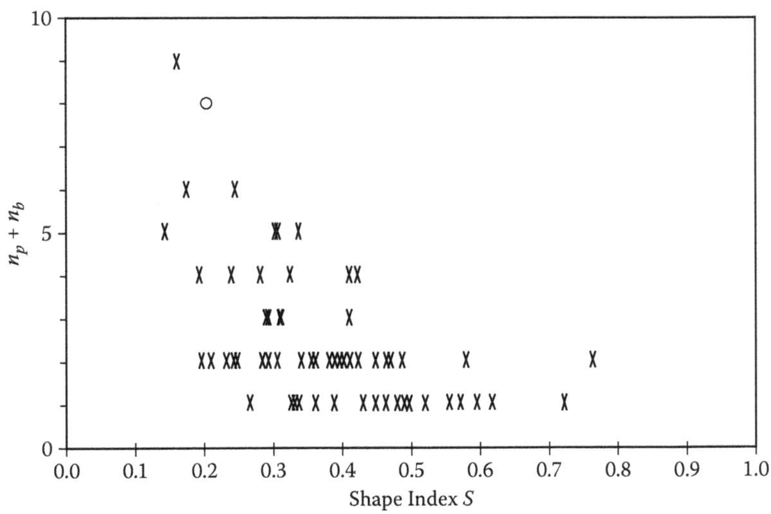

FIGURE 7.7 The relationship between shape index S and the number $n_p + n_b$ of peaks and bottoms of the employee distribution graph G_{EMP}.

one-peak and two-peak cities are about E = 200,000 and 300,000, respectively. As observed above, the employee distribution surface tends to become more complex as the city size increases in terms of population or employment.

7.3.3.3 Shape Index and Separation Number

In Figure 7.7, a negative linear relationship is observed and the number n_p of peaks is less than 3 whenever the shape index $S > 0.43$. Table 7.1a shows the average of shape index S and separation number N sorted by the number of peaks as well. The average shape index decreases monotonically with respect to the number of peaks of G_{EMP} (when $n_p = 1, 2, \ldots, 4$ and $n_p \geq 5$, the averages are 0.46, 0.37, 0.33, 0.30, and 0.24, respectively). The average separation number shows a not so monotonic increase (i.e., 2.4, 3.1, 3.6, 3.3, and 4.3).

7.3.4 Classification of Cities according to the Citywide Industrial Composition of Employment

One of the remaining factors of importance, which is related to the qualitative distribution of employees in a city, is its industrial composition of employment. To examine this relationship, a taxonomic classification of cities is conducted using the citywide composition of employees.

Specifically, we use Ward's method as a hierarchical clustering algorithm, whose superiority has already been reported with reference to ANOVA. The classification variables are the composition ratios of citywide employment of ten major industry groups,[5] and the dissimilarity metric is the Euclidean distance. The first-best result with four clusters is selected based on Beale's F value as a criterion to determine the number of clusters (Masuda, 1983).

As a result, the 65 cities are classified into four taxonomic *city groups*: I, II, III, and IV. Table 7.3 shows the average percentage industrial composition of each city group. From groups I to IV, the average shares of wholesale and retail trade and of services decrease monotonically; in contrast, that of manufacturing increases. This suggests that there are four distinct levels of service economy within a city: cities in group I, many of which are capital cities of prefecture, are at the highest level of specialization in the tertiary industry; cities in group IV are the typical manufacturing towns.

In Figure 7.2, the numbers of cities with the same (isomorphic) employee distribution graph G_{EMP} are also cross-tabulated by the city groups. All the monocentric cities, whose $n_p = 1$, belong to either group I or group II regardless of their population sizes and employment sizes. The upper limit of the number of peaks of G_{EMP} is 4 as long as a city is in group I. In contrast, cities belonging to group IV might have peaks of type Snd naturally (see Section 7.3.2). To be precise, we calculate the average number of peaks in each of the four city groups (the former decimal number in parentheses) and the ratio of the cities with one or more peaks of type Snd to the total number of cities in each group (the latter fractional number in parentheses) as group I (1.81, 3/14), group II (1.87, 12/33), group III (3.83, 11/13), and group IV (3.50, 5/5).

First, the average number of peaks of G_{EMP} differs significantly between group II and group III. On average, cities belonging to group I or II have two centers of employee distribution, while, in contrast, cities belonging to group III or IV have three or four centers.

Second, the ratio of the cities with peak(s) of type Snd increases monotonically as the city group changes from I to IV. When the employment share of manufacturing approaches the level of group III (nearly 30%), it is likely that one or more peaks will emerge as the spatially separated subcenter of manufacturing. (In group III, Okazaki and the connected twin cities of Numazu + Mishima are the only exceptional two-peak cities without peak of type Snd, whose peaks are types Trd and Mix.)

TABLE 7.3
Taxonomic City Groups of Citywide Composition of Employment

City Group (Number of Cities)	Average Composition Ratios of Employment by Major Industrial Groups (%)									
	Mining	Construction	Manufacturing	Wholesale and Retail Trade	Finance and Insurance	Real Estate	Transport and Communications	Electricity, Gas, Water, etc.	Services	Government Service
I (14)	0.12	10.49	9.94	34.42	4.82	1.38	8.58	0.92	22.88	6.01
II (33)	0.62	9.49	18.79	32.41	4.67	1.03	8.19	0.89	18.30	4.34
III (13)	0.06	7.82	29.82	27.70	4.07	0.84	7.84	0.83	16.78	4.00
IV (5)	0.09	6.55	46.31	21.86	2.24	0.49	5.61	0.62	11.96	1.97

From these figures, we notice on the one hand that a city specialized in the tertiary industry tends to show a rather simple and concentrated structure in employee distribution, and on the other that specialization in the secondary industry makes the structure more polycentric, accompanied by the functional differentiation of peaks.

7.4 CONCLUDING REMARKS

In this chapter, the qualitative structures of two-dimensional employee distribution in the 65 cities with population sizes larger than 100,000 were studied in Section 7.3 and compared to the earlier study of population distributions of 245 cities in Japan shown in Section 7.2. Using the graph theoretic method, the distributional characteristics were represented by the topological arrangement of local significant points on the surface: peaks, bottoms, and cols. According to the empirical study of employee and population distribution graphs G_{EMP} and G_{POP}, the following major results are obtained:

1. The graph theoretic method of Okabe (1981) is effective and efficient not only for the intercity comparison of the same distribution (i.e., population or employee alone) by isomorphism of graphs, but also for the comparison between different kinds of distribution (i.e., population and employee) in a city holding the two-dimensional characteristics of distributions.
2. In the case of the residential population distribution, the most frequently observed tree graphs G_{POP} are one-peak (monocentric) cities. The ratio of these cities to the total, 245, is two thirds, but there were only 12 monocentric cities among the 65 with population sizes greater than 100,000.
3. In the case of the workplace employee distribution among the 65 cities, the most frequently observed graph G_{EMP} is the two-node tree graph (of which there are 28, representing about 40%), that is, bipolar cities; the second most frequently observed is the one-node tree graph (of which there are 20, representing about 30%). In contrast, those of the population distribution graph G_{POP} are about 15% and 20%.
4. There is no tree graph of population distribution with more than seven peaks. In contrast, the maximum number of peaks of tree graph G_{EMP} is nine, and there exist more complex tree graphs G_{EMP} with more than four peaks.
5. The radial pattern without bottoms is very rare in population distributions, and the exceptional four cities show the three-way radial pattern. On the other hand, five employee distribution graphs G_{EMP} have a radial pattern and three of them (Shimonoseki, Hamamatsu, and Hiroshima) have a node of degree 4, which corresponds to the four-way radial pattern.
6. There is no circuit graph G_{EMP} except Kyoto (with one circuit) even if a city has a sizable population or total employment. This may suggest that inner city decline in employment is not yet so significant in Japanese mid-size cities.
7. On the whole, the distribution of the workplace employees shows a more simple and concentrated structure than that of residential population in

most of the 65 cities studied, in the sense that the employee distribution graph G_{EMP} has fewer nodes and circuits than the G_{POP} of a city.

8. In the case of the employee distribution, as the number n_p of peaks increases, distinctive gaps in the average population size are observed in between n_p = 2 and 3 (around P = 200,000) and n_p = 4 and 5 (around P = 400,000). And the upper limits of the population sizes of the one-peak and two-peak cities are about 300,000 and 550,000, respectively.
9. The shape of cities with the doughnut phenomenon of residential population is distinctively different from the shape of those without it. In the case of tree graphs of the employee and population distribution, on average, the more compact a city shape becomes, the fewer peaks G_{EMP} and G_{POP} of a city are observed.
10. Considering the four types of peak that represent the industrial composition of workplace around a peak, the original twelve isomorphic groups of G_{EMP} are segmented into 26 subgroups, most of which consist of only one city or precious few cities.
11. The 65 cities can be classified into four city groups using the citywide industry composition of employment, which suggests that there are four levels of service economy within a city. A city belonging to the group with a larger share of the tertiary industry tends to become a mono- or fewer-centric structure. In contrast, on average, when the industrial share of manufacturing in a whole city approaches 30%, one or more spatially separated subcenter(s) of manufacturing emerge, which leads to a more decentralized and polycentric structure.

Last, three remarks are appropriate. First, the basic spatial unit used in this study is a 1 km × 1 km quadrat, which might be too large to analyze small cities or to deal with more detailed roughness of the distribution surface. If possible, a smaller basic spatial unit would be desirable. Second, the method used above made it possible to compare the two-dimensional structures of different kinds of distribution, i.e., population and employment, and of different cities. This cross-sectional analysis might easily be expanded to the time-series analysis to examine a dynamic process of urban growth or decline. Third, it would be useful to deepen the descriptive analysis together with the theoretical studies of multicenter population distribution.

NOTES

1. One-dimensional analysis of population distribution $f(t)$ at a distance t from the CBD cannot deal with the two-dimensional characteristics of the density surface $f(x, y)$ at a location (x, y) on a geographical plane directly. On the other hand, trend surface analysis has difficulty in the comparison of different surfaces through their parameters. To overcome these shortcomings, alternative approaches to compare two-dimensional distributions were proposed by Warntz (1966), Fujii (1975), and so forth. The method used in this chapter is based on Okabe (1981) and is briefly summarized as follows (see also Chapter 2 of Okabe and Masuda, 1984).

1.1 Estimation of density surface: From point distribution, such as a dotted map, or smaller-area population statistics (i.e., census tract), it is possible to construct a smooth density surface (contour lines) by means of the histogram method or kernel estimation method. In the case of the 1 km × 1 km mesh data (standard grid square statistics), the surface is easily estimated by calculating the differences in density between adjacent quadrats.

1.2 Identification of singular points of local surface (peaks, bottoms, and cols): In a three-dimensional Cartesian space, the function $z = f(x, y)$, $(x, y) \in S$ forms a surface. If it is supposed that the surface is smooth and differentiable at any point (x, y) in closed domain S, the singular points where $V(x, y) = 0$ in the vector field $V(x, y) = (\partial f / \partial x, \partial f / \partial y)$ can be classified into three categories: peaks, bottoms, and cols. These categories are qualitative in the sense that each characteristic remains the same under any topological transformation with respect to S and any monotonically increasing function with respect to z.

1.3 Graphical expression of topological arrangement of peaks, bottoms, and cols: The integral curves of vector field $V(x, y)$ are the lines of steepest ascent of the surface. The graph whose nodes are peak points and whose links are the ridge lines connecting peak points represents the qualitative global structure of the surface.

1.4 Extraction of the significant structure by neglecting minor peaks and bottoms: The significant global structure is obtained by neglecting the peaks and bottoms whose relative heights and depths are less than a significant level Δ. In an empirical examination of population distribution and employee distribution, we set $\Delta = 1{,}500/\text{km}^2$. For reference, even if we set $\Delta = 1{,}000/\text{km}^2$ to include some miner (shallower) bottoms, there are only four cities whose employee distributions are represented by circuit graphs. These circuit graph cities are Sapporo, Kyoto, Hiroshima, and Kitakyushu, all of which are the ordinance-designated cities with larger population sizes P and employment sizes E than 700,000 and 400,000, respectively. This may suggest that inner city decline in employment (workplace population) is not yet so significant among mid-size cities in Japan.

2. This method is applicable not only in the intercity comparison among graphs $\{G_i\}$, but also in other comparisons, where the suffix i represents the different kind of distribution of a city (i.e., population in Section 7.2 and employment in Section 7.3) or time-series.
3. Note that this index is slightly different from the *compactness ratio* proposed by Richardson (1973), cited in Haggett et al. (1977). The constant is 4 because the basic spatial data unit is a grid square.
4. In the establishment census, persons engaged include not only all paid persons who are engaged in the business of establishments, but also individual proprietors and unpaid family workers. For the sake of simplicity, we use the term *employee* as a synonym for person engaged.
5. The industrial classification of an establishment is determined by the nature of the main business activities of the establishment according to the Standard Industrial Classification for Japan, and the ten major industry groups are as follows: (1) mining, (2) construction, (3) manufacturing, (4) wholesale and retail trade, (5) finance and insurance, (6) real estate, (7) transport and communications, (8) electricity, gas, heat supply, and water, (9) services, and (10) government service.
6. This relates to the combinatorial problem of the k-colored graph, whose nodes are labeled with different colors. Even if the case of the planer graph (as in the case of G_{EMP}) and k are fixed (as in these four peak types), as the number n of nodes increases, the combinatorial patterns rapidly increase.

REFERENCES

Administrative Management Agency of Japan. 1973. *Standard grid square and grid square code used for the statistics*. Announcement 143.

Clark, C. 1951. Urban population densities. *Journal of the Royal Statistical Society A* 114:490–96.

Fujii, A. 1975. The study of activity contour lines 1: Structural concepts of closed curve figures. *Transactions of the Architectural Institute of Japan* 267:121–27 (in Japanese).

Haggett, P., Cliff, A. D., and Frey, A. 1977. *Locational analysis in human geography*. London: Arnold.

Japanese Bureau of Statistics. 1977. *1975 population census of Japan*. Vol. 1. Tokyo: Office of the Prime Minister.

Masuda, S. 1983. Cluster analysis on the structure of employees distributions of eight major cities in Tokyo metropolitan area. *Papers of the Annual Conference of the City Planning Institute of Japan* 18:55–60 (in Japanese).

Masuda, S. 1987. Development of GIS for urban analysis and its application to regional clustering. PhD dissertation, Department of Urban Engineering, University of Tokyo (in Japanese).

Okabe, A. 1981. *A qualitative method of trend surface analysis*. Discussion Paper 3, Department of Urban Engineering, University of Tokyo.

Okabe, A., and Masuda, S. 1984. Qualitative analysis of two-dimensional urban population distributions in Japan. *Geographical Analysis* 16:301–12.

Warntz, W. 1966. The topology of a socioeconomic terrain and spatial flows. *Papers, Regional Science Association* 17:47–61.

8 An Empirical Analysis of Consumers' Evaluation of Department Stores

Ikuho Yamada and Yukio Sadahiro

8.1 INTRODUCTION

Understanding consumers' spatial choice behaviors has been an important research focus in the field of regional science. Shopping environments where consumers exercise their choices change continuously, affecting as well as being affected by such choices and preferences. For instance, the large-scale retail market in Japan was first dominated by department stores that emerged at the beginning of the westernization of Japan (i.e., the end of its national isolation policy) around 1900; the department stores were the symbol of metropolises and the most fashionable, popular choice for urban consumers to shop in (Sugioka, 1991; Takaoka and Koyama, 1991). During the period of rapid economic growth after World War II, however, chain stores of general supermarkets that offered much lower prices than department stores came to dominate the Japanese retail market, especially in urban areas. Further, the motorization in Japan that followed this development generated the need for a new shopping environment to support access by car with large parking facilities. Accordingly, new types of retailers such as roadside stores, specialty chain stores, suburban-style department stores, and shopping mall complexes have burgeoned in recent years, resulting in intensified competition between traditional retailers, including urban department stores and those newly developed large-scale retailers. As the shopping environments increasingly diversify spatially, consumers' spatial choice behaviors are also becoming so complex that it is necessary to investigate them under these various settings.

The majority of literature on consumers' behaviors in large-scale retail facilities focuses on general supermarkets and shopping malls (Bearden, 1977; Bellenger et al., 1977; Sasaki et al., 1980; Honda, 1983; Kondo and Aoyama, 1989; Asada et al., 1991), and relatively little has been investigated in the context of department stores and other large-scale stores (Heidingsfield, 1949; Rich, 1963; Hirschman, 1979). Department stores, however, have one of the largest market areas in Japanese cities, and they are thus expected to have strong influence upon the structure of the retail market. Consumers' spatial choice behaviors in the context of department stores thus warrant more attention and investigation.

Behavior models such as Huff's (1963), gravity (Stewart, 1948), and logit (Ben-Akiva and Lerman, 1985) have often been used in the literature of this field (Koide, 1976; Weisbrod et al., 1984). Those models attempt to describe consumers' spatial choice behaviors by objective measures of retailers' characteristics (e.g., square footage, price), consumers' characteristics (e.g., age, gender), and spatial relationships between them (e.g., distance between consumers' residence and retailers). On the other hand, when consumers choose a retail store in reality, their choice is likely to be affected by ambiguous, subjective information, such as their past experience at a given retailer and its reputation. In addition, it is impossible for consumers to recognize those objective elements of retailers accurately and comprehensively, so that the knowledge or understanding of each retailer should vary both quantitatively and qualitatively between individuals.

Shopping at department stores is mostly associated with higher-level consumer goods such as luxury clothes and durable consumer goods, which correspond to *shopping goods* defined by Copeland (1923, p. 283) as "those for which the consumer desires to compare prices, quality, and style at the time of purchase." Consumers often visit multiple stores when purchasing those goods to make a comparison between stores. The subjective, psychological elements affecting consumers' choice behaviors, which can be seen as personal preference or image of retailers, can be expected to play a particularly important role in the context of shopping at department stores, including the decision about which stores to shop in. The marketing strategy of department stores also takes into account such psychological, personal effects, unlike general supermarket chains, which focus on lower prices. For example, the former emphasizes provision of a broad selection of merchandise for target consumer groups, interior and exterior design of stores, quality and variety of available services, and cultural, nonsales events (e.g., art exhibition). Given the emphasis that the department stores' marketing strategy places on the psychological, personal elements, understanding what constitutes them and how they are related to one another is the important first step toward effective analysis of consumers' choice behaviors regarding department stores.

In light of the discussion above, this study aims to explore image or preference that individuals may have regarding department stores, which will be referred to as *evaluation* or *image evaluation* hereinafter, and to further analyze the mechanism that forms such personal evaluation of retailers. To obtain direct and concrete information about individuals' evaluation of department stores, a questionnaire survey was conducted in 1996. Results of the survey are first examined by a nonparametric method proposed in this chapter to derive "average" or general evaluation among a specific group of consumers who participated in the survey. The analysis of the evaluation mechanism is then carried out with rank logit models.

The next section summarizes basic information about the questionnaire survey, such as its objectives, questions, and respondents. The section that follows proposes a nonparametric method to derive average evaluation from ranking data. It also discusses the results of the method's application to the rankings of department stores provided by the survey respondents. The fourth section briefly describes the rank logit model first and then explains how it is implemented to examine the mechanism of consumers' evaluation of department stores. Based on the implemented models, the next two sections analyze relationships between department stores' evaluation and their characteristics, and the relationships between the evaluation and characteristics of individual consumers, respectively. Validity of the derived relationships is examined through comparison with the average evaluation obtained by the nonparametric method. The final section provides conclusions and discusses future research questions.

8.2 QUESTIONNAIRE SURVEY

8.2.1 Survey Outline

As stated above, this study hypothesizes that the subjective evaluation (or image) that individual consumers have of each department store, in addition to objective or physical characteristics of each store, will affect their choice of department stores to shop in. As a first step toward understanding consumers' evaluation of individual department stores and its influence on their choice of stores, a questionnaire survey was carried out to collect direct and concrete information about consumers and their use and image evaluation of department stores.

The survey was targeted at female homemakers aged 30 to 50 who resided in the 23 special wards or in the western district of Tokyo, Japan. We decided to focus on female homemakers mainly because they could be assumed to utilize department stores most frequently among members of a household. In addition, male and female consumers may have different store evaluations and psychological mechanisms for the evaluation; our limited budget in terms of both money and time did not allow us to collect samples large enough to take into account such potential gender differences. Because a large number of department stores exist in Tokyo, it is reasonable to assume that individual consumers shop in or are familiar with only a limited number of them, which is likely to include those that are easily accessible from their homes. The survey target area was therefore further limited to residential areas along three major railway lines connecting the center of Tokyo and its bedroom communities, so that respondents' familiarity would cover at least partially similar sets of department stores. The three railway lines were the Odakyu, Tokyu-den'entoshi, and Tobu-tojo lines.

A principal survey was conducted during October and early November 1996, followed by a supplementary one in December to adjust unbalanced spatial distribution of respondents.

8.2.2 Survey Questions

In this study we postulate that consumers' evaluation of department stores is affected not only by the objective and observable characteristics of the stores (e.g., square footage, the number of years since the store was opened, marketing strategy) but also by consumers' socioeconomic and demographic characteristics and their relationships with each store (e.g., access time, ease of access by train—that is, whether or not the store and a consumer's residence are directly connected by a railway line—, and familiarity). We further hypothesize that the evaluation that an individual has for a given department store is related to how often she would visit the store. The survey therefore consists of the following three categories of questions: personal profile, frequency of shopping activities, and subjective image that respondents have of individual department stores. Each category is briefly explained below.

8.2.2.1 Personal Profile

As socioeconomic characteristics may affect individuals' shopping frequency and store choices, respondents were asked to provide information about their age, household income, possession of a driver's license, employment status, and household size and structure. Since residential location and accessible railway lines would also have an influence on individuals' choice of department stores, questions on their residential address, the closest railway station, and transportation mode to access the station were included, too.

8.2.2.2 Frequency of Shopping Activities

Generally speaking, how often a consumer visits a specific department store relates to the frequency of her visits to department stores as a whole, which itself further relates to the frequency of her shopping trips, including those to grocery stores and other superstores (note that Japanese department stores sell groceries, unlike the ones in the United States). Respondents were therefore asked to provide the average number of shopping trips per month, that of visits to any department stores, and a list of the ten department stores that they most frequently visited, together with the number of visits to each store.

8.2.2.3 Subjective Image of Individual Department Stores

Respondents were asked to rank a set of nine department stores, listed in Table 8.1, in terms of ten image items, each representing a certain aspect of service that department stores offer.

Each of the ten image items was selected according to two criteria: (1) it can be reasonably assumed that individuals have relatively fixed ideas about that aspect of department stores, and (2) that aspect is related to sales promotion of department stores that can be distinguished from that of other large-scale retailers, such as supermarkets. The image items can be grouped into three with respect to their emphasis. *Quality* items refer to those that are related to quality of service. *Fashion* items refer to those representing how well a department store catches up with a trend of the time. *Usability* items refer to those representing how comfortable a store is to shop in. The ten image items are listed in Table 8.2.

To select the nine target department stores, the following four criteria were considered. First, the selected set of stores must cover the central commercial areas of Tokyo, that is, Nihonbashi, Ginza, Shinjuku, Shibuya, and Ikebukuro. Second, the stores must be located in the 23 special wards of Tokyo. Third, they must be widely recognized. Last, the selection must have a good mixture of different department store companies. An exception to the second criterion is Takashimaya in Futakotamagawa, which is located in the survey area. This store was included because of its relatively large size as a department store located outside the 23 special wards. During the survey period, a new large department store, Takashimaya

TABLE 8.1
Major Department Stores in Tokyo Selected as Survey Targets

Company Name	Location
Takashimaya	Nihonbashi
Mitsukoshi	Nihonbashi
Matsuzakaya	Ginza
Isetan	Shinjuku
Odakyu	Shinjuku
Seibu	Shibuya
Tokyu honten[a]	Shibuya
Tobu	Ikebukuro
Takashimaya	Futakotamagawa

[a] Honten translates from the Japanese as the "head store" of a company. Because there are multiple Tokyu department stores in Shibuya, the store under consideration is distinguished by that term.

TABLE 8.2
Image Items and Their Categories

Quality Items (5)	Fashion Items (3)	Usability Items (2)
Luxuriousness	Fashionableness/urbaneness	User-friendliness
Breadth of merchandise selection	Sensitivity to trends	Comfortable atmosphere for shopping
Service quality of salespersons	Youthfulness/freshness	
Elegant atmosphere		
Attractiveness of nonsales events		

in Shinjuku, was opened. We decided, however, to exclude it from our survey, primarily because not many of the respondents would have a chance to visit the store before completing the questionnaire. It was also suspected that its intensive and extensive opening campaign might bias respondents' evaluation. Ideally, multiple department stores for each commercial area, as well as for each department store company, should be selected to enable a more comprehensive comparison. We nonetheless decided to limit this survey to the nine stores, because including more stores would make the questionnaire so complex that respondents might be confused or discouraged from answering it.

8.2.3 Basic Summary of Respondents' Profiles

One hundred and nine effective questionnaires were obtained from 178 that were sent out during the survey period (effective response rate of 61.2%). The average age of the 109 respondents was 45.5, and 59% of them were employed at the time of the survey. The average frequencies of general shopping trips and of trips to department stores were 4.17 and 3.16 per week, respectively. The income distribution and the distribution of driver's license possession are summarized in Table 8.3.

TABLE 8.3
Income and Driver's License Possession of Survey Respondents

Income	(In Dollars)	%
Above 10 million yen	(>100K)	49
8~10 million yen	(80~100K)	26
6~8 million yen	(60~80K)	19
4~6 million yen	(40~60K)	6
Total		100
Driver's License		**%**
Car only		68
Car and motorbike		27
None		5
Total		100

8.3 AVERAGE EVALUATION OF DEPARTMENT STORES

8.3.1 METHOD

This section discusses the derivation of average evaluation of the nine department stores with respect to each of the ten image items using the ranking data provided by the survey respondents. Because the rankings are ordinal-scale data, and a value given to a specific department store only indicates its relative superiority or inferiority in comparison with other stores, ordinary measures of central tendency, such as the arithmetic mean, do not provide a meaningful summary of the rankings since they are designed for interval- and ratio-scale data (Burt and Barber, 1996). To analyze relationships among such ranking data, several variants of rank correlation coefficients—namely, Goodman-Kruskal's γ, Kendall's τ, and Spearman's r_s—have been proposed (Upton, 1978). Nishisato's dual scaling (1982) can also be used to quantify qualitative data including the rankings. In addition, it is known that Luce's choice axiom (1959) is applicable to the rankings, and thus the stochastic choice theory can also be applied (Fararo, 1978). From these characteristics of ranking data, this study assumes that a sequence of the rankings that is most likely expectable from the rankings given by the survey respondents can be seen as a reliable average or representative evaluation of the nine department stores. We therefore propose the following method to obtain the average rankings from individual respondents' ranking information.

Let D_{ij} be the rank that individual j gives to department store i and D_i be the average rank of i that summarizes ranks given by all individuals. Further denote the number of individuals who rank department store i higher than department store k (that is, $D_{ij} > D_{kj}$) by $N(i \succ k)$ and the number of individuals who rank department store i lower than department store k (that is, $D_{kj} > D_{ij}$) by $N(k \succ i)$. Note that individuals who rank department stores i and k equally are excluded from these numbers. It should also be mentioned that not all survey respondents ranked all of nine department stores, so that the sum of $N(i \succ k)$ and $N(k \succ i)$ represents the number of respondents who ranked both department stores i and k, excluding those who ranked them equally.

First, a pair of department stores i and k that maximizes

$$\frac{\max\left(N(i \succ k), N(k \succ i)\right)}{N(i \succ k) + N(k \succ i)}$$

is searched for. If $N(i \succ k) > N(k \succ i)$, for example, it is considered that relative evaluation comparing department stores i and k is $D_i > D_k$ on average. The second step is to search for a pair that gives the second largest value of

$$\frac{\max\left(N(i \succ k), N(k \succ i)\right)}{N(i \succ k) + N(k \succ i)}$$

and then determine their relative evaluation in the same way. This step is repeated until obtained relative evaluations contradict one another. For instance, if the first three relative evaluations are $D_1 > D_2$, $D_2 > D_3$, and $D_3 > D_1$, in this order, the third evaluation (i.e., $D_3 > D_1$) contradicts the first two so that the procedure is aborted here. Then $D_1 > D_2 > D_3$, obtained by combining the first two evaluations, is used as the average rankings of the department stores. As mentioned earlier, this method does not consider ties while focusing on evident differences in individuals' evaluation of the department stores. Effective treatment of ties to improve the validity of the obtained average rankings will require further investigation.

8.3.2 Results

Table 8.4 shows the average rankings of the nine department stores obtained by the method explained above. Based on these rankings, this section looks into characteristics of department stores that potentially influence the evaluation associated with each image item.

Let us examine the quality items first. For all five image items belonging to this category, the two department stores located in Nihonbashi (Takashimaya and Mitsukoshi) are ranked among the highest, which is a distinctive characteristic of this category in comparison with the other two. Given that Nihonbashi has a long history as a central commercial area in Tokyo, and that the two department stores are among the oldest department stores in Japan, it appears that long-established tradition in terms of both location and department stores themselves is an important element in forming consumers' evaluation of the quality of department stores. As for the fashion items, location does not seem to have much influence. For example, while both Shibuya and Ikebukuro are often referred to as "towns for young people," there is no clear pattern in the rankings of the department stores located there. Generic images associated with individual department stores may therefore be expected to influence the image items in this category. Finally, the usability items are expected to be closely related to individuals' personal characteristics by definition. It is thus likely that relationships between individuals and department stores, such as the frequency of visits and access time or distance, would also have a strong influence on the image items in this category.

To assess the hypothesis that locational relationships between consumers and department stores impact on evaluations that individual consumers have for each store, we further examined regional variations in the average rankings. The respondents were split into three groups, Shinjuku, Shibuya, and Ikebukuro, according to the terminal stations of the railway lines they used. The method explained in Section 8.3.1 was applied to each group to obtain the average rankings specific to the region. Figure 8.1 summarizes the resulting average rankings (please note that only selected image items are presented here owing to space limitations).

The resulting regional average suggests that "attractiveness of nonsales event," "comfortable atmosphere for shopping," and "user-friendliness" have the largest regional variations in the evaluation; the variations are particularly large for the first two image items. For these image items, department stores located at the terminal station of a specific region or along the associated railway line tend to be given higher evaluations. This tendency also holds for image items with intermediate regional variations; these image items are "service quality of salespersons" and "breadth of merchandise selection." On the other hand, image items that have small regional variations are "luxuriousness," "elegant atmosphere," "fashionableness/urbaneness," "sensitivity to trends," and "youthfulness/freshness." The last diagram in Figure 8.1 shows that there is little variation in the ranking evaluation of "youthfulness/freshness." This result implies that, for these five image items, a relatively well-established consensus of evaluation of department stores exists among consumers regardless of their residential location.

The results above have led to three important findings about consumers' image evaluation of department stores. First, the image items that are rather conceptual (e.g., "luxuriousness," "youthfulness/freshness") appear to have a relatively established evaluation on each department store, and it does not vary much according to consumers' residential location. Second, the image items that are rather concrete and related to actual shopping experience in stores (e.g., "attractiveness of nonsales event," "comfortable atmosphere for shopping") have large regional variations in the evaluation of individual stores. In relation to this, the third finding is that, for the items with large regional variations, evaluation tends to be higher if department stores are located along the railway line that a given consumer uses or at its terminal station, which may imply frequent visits to the stores compared with other stores. In addition, if we focus on the department stores rather than the image items, two tendencies can be pointed out. First, department stores that hold a large market area, such as Takashimaya in Nohonbashi and Isetan in Shinjuku, generally have high evaluation without substantial regional variations. On the other hand, evaluation of department stores that have a smaller market area and are located along the railway line or at its terminal station tend to be strongly affected by whether or not a consumer uses the railway line.

From these findings, we hypothesized the mechanism of consumers' evaluation of department stores as below. Assuming that the evaluation that an individual has for a given department store can be decomposed into two elements, one that is common to all consumers and one that varies between individuals, the relative importance of

TABLE 8.4
Average Rankings of the Nine Department Stores for the Ten Image Items

Category	Image Item	1	2	3	4	5	6	7	8	9
Quality	Luxuriousness	Takashimaya (N)	Mitsukoshi (N)	Isetan (Sj)	Tokyu (Sb)	Takashimaya (F)	Matsuzakaya (G)	Seibu (Sb)	Odakyu (Sj)	Tobu (I)
	Breadth of merchandise selection	Isetan (Sj)	Takashimaya (N)	Mitsukoshi (N)	Seibu (Sb)	Tobu (I)	Odakyu (Sj)	Tokyu (Sb)	Matsuzakaya (G)	Takashimaya (F)
	Service quality of salespersons	Takashimaya (N)	Mitsukoshi (N)	Isetan (Sj)	Matsuzakaya (G)	Tokyu (Sb)	Takashimaya (F)	Odakyu (Sj)	Seibu (Sb)	Tobu (I)
	Elegant atmosphere	Mitsukoshi (N)	Takashimaya (N)	Tokyu (Sb)	Matsuzakaya (G)	Isetan (Sj)	Takashimaya (F)	Odakyu (Sj)	Seibu (Sb)	Tobu (I)
	Attractiveness of nonsales events	Mitsukoshi (N)	Takashimaya (N)	Isetan (Sj)	Tokyu (Sb)	Matsuzakaya (G)	Odakyu (Sj)	Seibu (Sb)	Tobu (I) Takashimaya (F)	
Fashion	Fashionableness/ urbaneness	Isetan (Sj)	Seibu (Sb)	Takashimaya (F)	Takashimaya (N)	Mitsukoshi (N) Tokyu (Sb)		Odakyu (Sj)	Matsuzakaya (G)	Tobu (I)
	Sensitivity to trends	Isetan (Sj)	Seibu (Sb)	Takashimaya (F)	Tokyu (Sb)	Odakyu (Sj)	Tobu (I)	Takashimaya (N)	Mitsukoshi (N)	Matsuzakaya (G)
	Youthfulness/ freshness	Isetan (Sj)	Seibu (Sb)	Takashimaya (F)	Odakyu (Sj)	Tobu (I)	Tokyu (Sb)	Matsuzakaya (G)	Takashimaya (N)	Mitsukoshi (N)
Usability	User-friendliness	Isetan (Sj)	Odakyu (Sj)	Takashimaya (F)	Seibu (Sb)	Tokyu (Sb)	Takashimaya (N)	Tobu (I)	Mitsukoshi (N)	Matsuzakaya (G)
	Comfortable atmosphere for shopping	Isetan (Sj)	Odakyu (Sj)	Takashimaya (F)	Seibu (Sb)	Takashimaya (N)	Mitsukoshi (N)	Tokyu (Sb)	Matsuzakaya (G)	Tobu (I)

Note: Characters in parentheses represent store locations: N = Nihonbashi; G = Ginza; Sj = Shinjuku; Sb = Shibuya; I = Ikebukuro; F = Futakotamagawa.

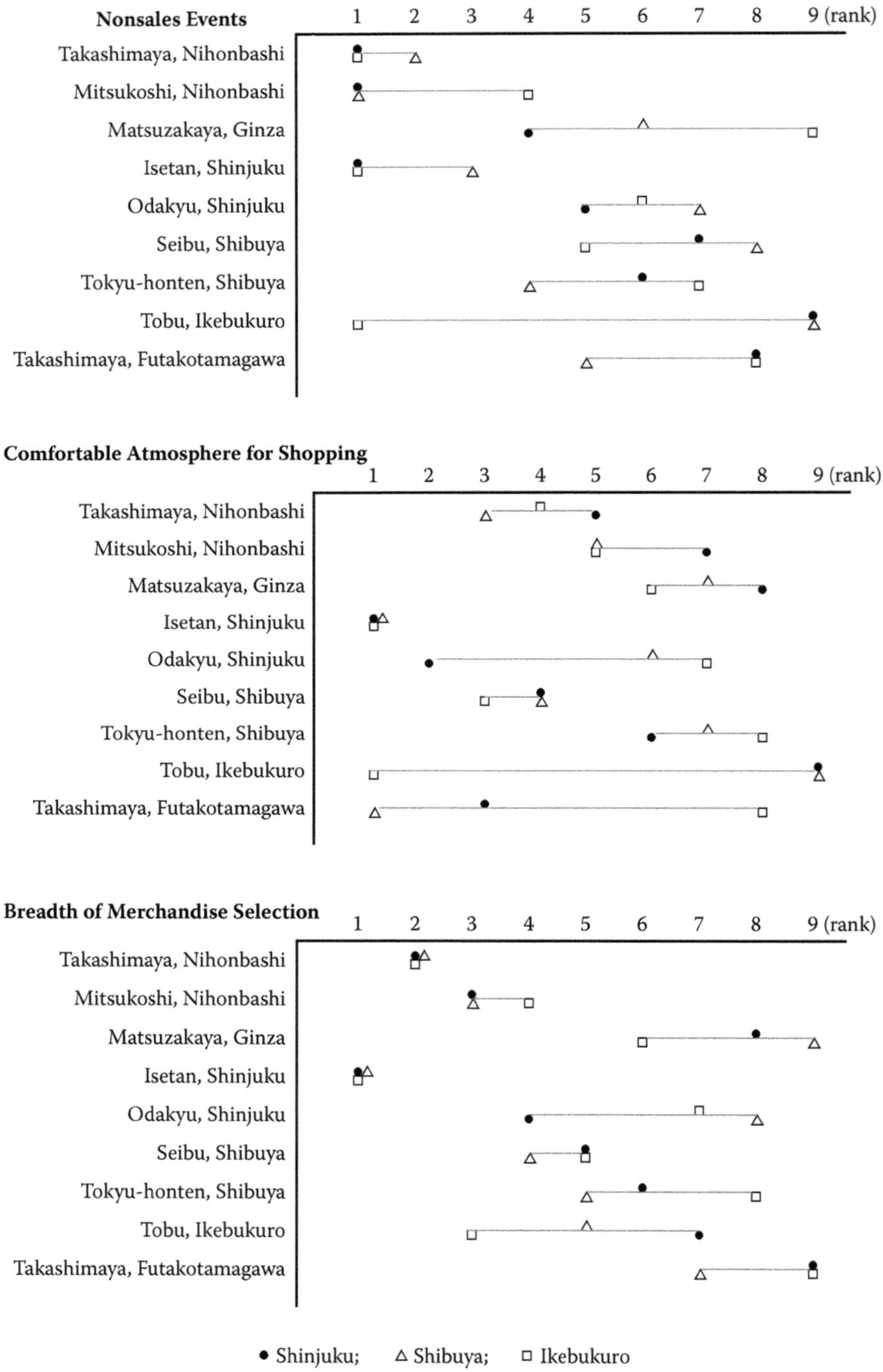

FIGURE 8.1 Regional differences in the rankings of the nine department stores for selected image items.

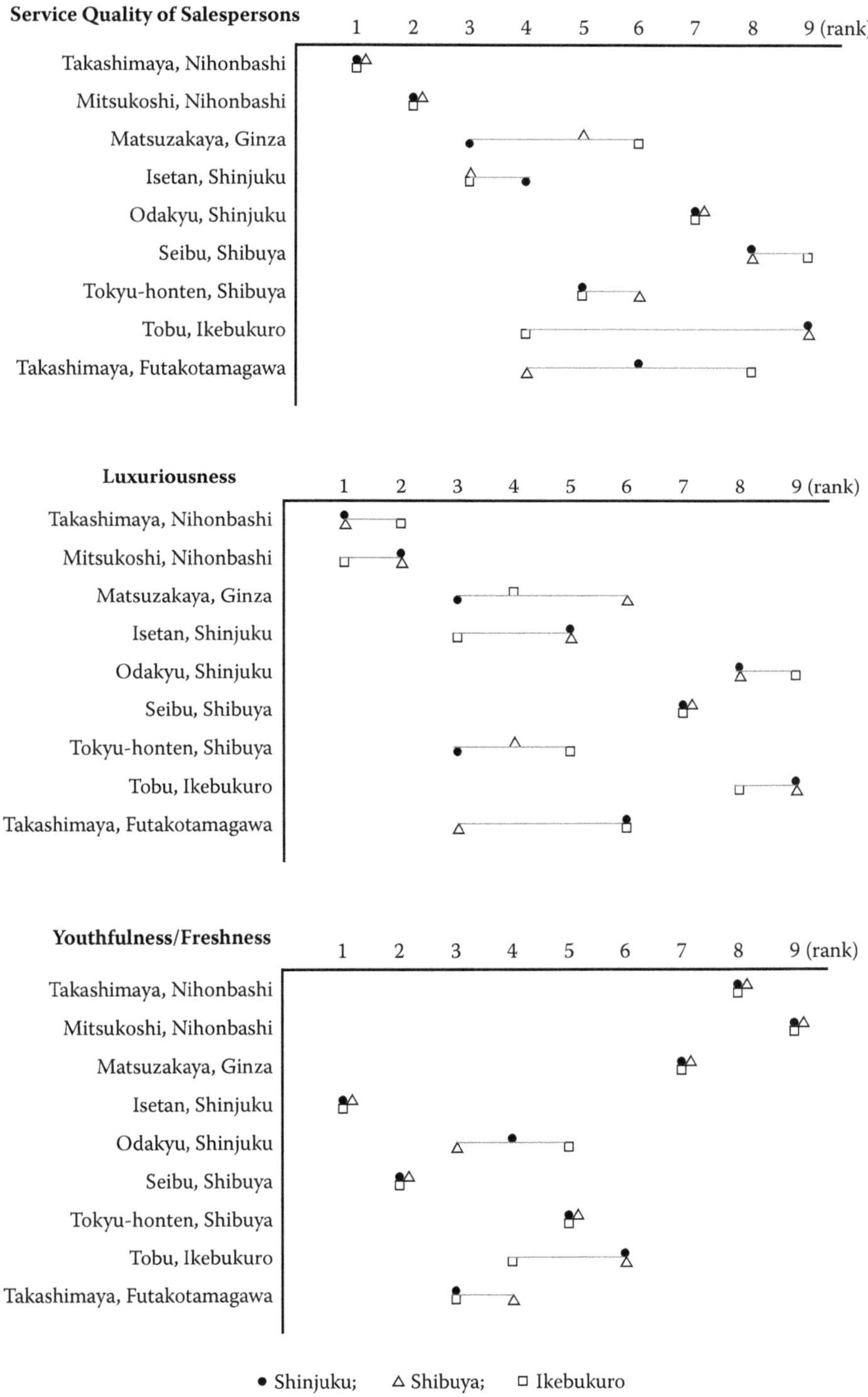

FIGURE 8.1 *(Continued)*

the two elements considerably varies between the image items. Further, it is hypothesized that the former element can be explained by characteristics of department stores themselves, such as square footage, location, tradition, and merchandise selection, and the latter can be explained by characteristics of individuals and their relationships with stores, such as accessibility and familiarity. The next three sections discuss quantitative modeling of the evaluation mechanism using these hypotheses.

8.4 DEPARTMENT STORE IMAGE EVALUATION MODELING

This and the next two sections are devoted to quantitative modeling of the evaluation mechanism that consumers have for department stores based on the hypotheses developed in the previous section. This section first provides the basics of the rank logit model and then explains its application to the modeling of consumers' image evaluation of department stores.

8.4.1 Basics of the Rank Logit Model

Discrete choice models are used to model individuals' choice from a finite set of mutually exclusive alternatives (Ben-Akiva and Lerman, 1985), for example, individuals' choice of a department store to shop in from a set of department stores that can be reached in a reasonable time. The logit model is a type of such discrete choice models, and the rank logit model utilized in this study is its extension for handling ranking data (Japan Society of Traffic Engineers, 1993).

On the basis of the random utility theory (Domencich and McFadden, 1975), discrete choice models in general assume that the utility of alternative i for individual j, U_{ij}, consists of a measurable or systematic component, V_{ij}, that is determined as a function of measured attributes of the individual and his or her alternatives and a random component, e_{ij}, that is caused by any immeasurable factors such as idiosyncrasies of each individual and errors in data. That is,

$$U_{ij} = V_{ij} + e_{ij} \tag{8.1}$$

The random utility theory further postulates that individuals always select an alternative that maximizes their utility. The logit model is a special case of this general discrete choice model where the random component e_{ij} is assumed to follow the Gumbel distribution. Let K_j be a set of alternatives available to individual j. The probability that individual j chooses alternative i, P_{ij}, equals the probability that the utility of i is larger than any other alternative in K_j and can therefore be represented as below:

$$P_{ij} = \Pr\ U_{ij} > U_{kj},\quad k \in K_j \tag{8.2}$$

$$P_{ij} = \exp\left(V_{ij}\right) \Big/ \sum_{k \in K_j} \exp\left(V_{kj}\right) \tag{8.3}$$

Further, the systematic component V_{ij} of the utility U_{ij} is generally represented as a linear combination of attributes of alternative i and individual j. Letting S be the total number of attributes included in the linear model, Z_{ijl} be the lth of the S attributes, and β_l be a parameter associated with the lth attribute, V_{ij} is written as

$$V_{ij} = \sum_{l=1}^{S} \beta_l Z_{ijl} \tag{8.4}$$

In the rank logit model, the rankings are hypothesized to be a series of choices. The rankings of K alternatives are decomposed into $(K - 1)$ choices that are statistically independent of one another. The alternative that is ranked first is the most desirable among the K alternatives, the one that is ranked second is the most desirable among the $K - 1$ alternatives, excluding the first one, and so on. The process continues until $(K - 1)$ choices are obtained. Assuming that the random component e_{ij} of individual alternatives is distributed independently, the probability that the rankings made by individual j follow a particular order can be represented as a product of probabilities associated with the $(K - 1)$ choices. Let us denote the probability that alternative k is chosen from an alternative set $\{k, k+1, \cdots, K\}$ by $P_j\left(k \middle| \{k, k+1, \cdots, K\}\right)$. This probability is given by

$$P_j\left(k \middle| \{k, k+1, \cdots, K\}\right) = \exp\left(V_{kj}\right) \Big/ \sum_{i=k}^{K} \exp\left(V_{ij}\right) \tag{8.5}$$

The probability that alternative 1 is ranked first, alternative 2 is ranked second, ..., and alternative K is ranked Kth, denoted by $P_j(1, 2, \cdots, K)$, is then expressed as

$$\begin{aligned} P_j(1,2,\cdots,K) &= P_j\left(1 \middle| \{1,2,\cdots,K\}\right) P_j\left(2 \middle| \{2,3,\cdots,K\}\right) \cdots P_j\left(K-1 \middle| \{K-1,K\}\right) \\ &= \prod_{k=1}^{K-1} P_j\left(k \middle| \{k,k+1,\cdots,K\}\right). \end{aligned} \tag{8.6}$$

The log-likelihood function L for the rankings of K alternatives by J individuals is a function of the parameters $\beta_l(l = 1, \cdots, S)$, and the maximum likelihood estimation method can be used to estimate β_l if ranking data are available (for details see Japan Society of Traffic Engineers, 1993).

Since the rank logit model is a type of disaggregate behavior models, it inherits one of their characteristics: because they have high data efficiency owing to the direct use of individual data, they can generate a model that explicitly describes individuals' behavior from relatively small data. This characteristic of the rank logit model makes it suitable for the present study, which deals with subjective data on individuals' rankings of department stores and aims to create individual-based models with a variety of explanatory variables, including attributes of both department stores and individuals using a relatively small dataset.

8.4.2 Modeling of Consumers' Evaluation of Department Stores

In this study, we create department store image evaluation models to describe the mechanism of individuals' evaluation, in particular, rankings of department stores with respect to the ten image items using the rank logit model framework.

We here assume that the image evaluation of department store i by individual j corresponds to the utility that the department store offers the individual, U_{ij}, and individual j ranks department store 1 higher than department store 2 if $U_{1j} > U_{2j}$. The systematic component V_{ij} of U_{ij} is formulated as a linear combination of the following three groups of explanatory variables.

8.4.2.1 Variables Associated with Department Store *i*, X_{ik}

As characteristics of department stores may influence consumers' evaluation of them, this study focuses on merchandise (quality and selection), interior and exterior design of stores, nonsales events, and established tradition. More specifically, the following five variables are used, each corresponding to these characteristics: square footage of a store's shopping area (X_{i1}), the number of designer-brand tenants (X_{i2}), the number of years from the most recent renovation of the store (X_{i3}), the number of nonsales events per year (X_{i4}), and the number of years since the store was opened (X_{i5}). In addition, dummy variables ($X_{i6} \sim X_{i13}$) indicating each department store are also included in the models to capture established or fixed evaluation that is specific to the store. Among the 13 department store variables, correlation between the square footage X_{i1} and the number of designer-brand tenants X_{i2} is rather high, and caution should be exercised in interpretation of estimated models.

8.4.2.2 Variables Associated with Individual Consumer *j*, Y_{jl}

The analysis discussed in Section 8.3 showed that an individual's evaluation of department stores varies depending on the terminal station that she uses. Dummy variables, Y_{j1} and Y_{j2}, indicating the terminal stations of Shinjuku, Shibuya, and Ikebukuro, are therefore added to the models. According to Morichi and colleagues (1984), quality, store types, and trip characteristics, such as modes of transportation, have much greater influence on individuals' choice of commercial areas than personal demographic characteristics, while household income has a relatively stronger influence on shopping trips involving longer trip distance. In the light of this, household income (Y_{j3}) is also included in the models.

8.4.2.3 Variables Representing Relationships between Department Store *i* and Individual *j*, Z_{ijm}

Morichi and colleagues (1984) also have found that consumers' evaluation of commercial areas is markedly affected by familiarity, convenience, and price, in that order. In the discussion in Section 8.3, there is another finding that individuals tend to highly evaluate department stores that are located along the railway line that they use or at its terminal station. Given these findings, two variables are included in the models to capture relationships between individuals and department stores. One is travel time by rail from a train station closest to individuals' residence to department stores (Z_{ij1}), and the other is frequency of visits (Z_{ij2}) as a proxy of individuals' familiarity with stores.

The department store image evaluation models developed in this study represent image evaluation U_{ij} of department store i by individual j as a linear combination of a systematic component V_{ij} and a random component e_{ij}, as shown in Equation (8.1). The systematic component V_{ij} is formulated by use of a certain combination of the 18 variables described above. For example, when all variables are used, the systematic component V_{ij} is given by

$$V_{ij} = \sum_{k=1}^{13} \alpha_k X_{ik} + \sum_{l=1}^{3} \beta_l Y_{jl} + \sum_{m=1}^{2} \gamma_m Z_{ijm} \tag{8.7}$$

where α_k ($k = 1, \ldots, 13$), β_l ($l = 1, 2, 3$), and γ_m ($m = 1, 2$) are model parameters.

While we have investigated a variety of model structures with different combinations of the variables, this chapter hereinafter focuses on two structures that have been found to have higher explanatory power than others, and thus can be expected to better reflect consumers' evaluation mechanism. The next two sections discuss estimation of the department store image evaluation models for each image item by means of the maximum likelihood estimation method. Section 8.5 deals with models consisting of the department store variables (X_{ik}), and Section 8.6 extends them to include the variables representing relationships between individuals and stores (Z_{ijm}).

Akaike information criterion (AIC) (Sakamoto et al., 1983) is used to assess the significance of estimated models. AIC is defined as below, and a model that minimizes the AIC value is considered to be the best model:

$$\begin{aligned} \text{AIC} = &-2\left(\text{the maximum log-likelihood of the model}\right) \\ &+ 2\left(\text{the number of parameters in the model}\right) \end{aligned} \tag{8.8}$$

This definition of AIC implies that, if multiple models have a similar level of the maximum log-likelihood, a model with the fewest parameters should be selected.

8.5 RELATIONSHIPS BETWEEN DEPARTMENT STORE CHARACTERISTICS AND THEIR IMAGE EVALUATION

8.5.1 Model Development

The analysis in Section 8.3 has led to the following hypotheses, assuming that evaluation of a given department store by an individual can be decomposed into an element that is common to all consumers and another that varies between individuals. First, the two elements have considerably different relative importance depending on the image items. Second, the common element can be explained by characteristics of department stores, while the individually varying element can be associated with relationships between individuals and stores. This section discusses the department image evaluation models that focus on the common element. Three separate models are developed using different structures for the systematic component V_{ij} of the individual's image evaluation U_{ij}.

8.5.1.1 Constant Model

The first model attempts to express the component V_{ij} of the image evaluation that is common to all consumers directly by a set of constants representing fixed evaluation already established for individual department stores. In other words, this model only includes the department store dummy variables ($X_{i6} \sim X_{i13}$), that is,

$$U_{ij} = \sum_{k=6}^{13} \alpha_k X_{ik} + e_{ij} \tag{8.9}$$

and parameters estimated for the dummy variables ($\alpha_6 \sim \alpha_{13}$) represent fixed, established evaluation of the department stores. The model is called a constant model.

8.5.1.2 Attribute Model

The second model attempts to capture the common component V_{ij} by measurable, quantitative attributes of department stores and is called an attribute model. This model is to investigate which attributes of department stores impact on their evaluation for a particular image item and is formulated as

$$U_{ij} = \sum_{k=1}^{5} \alpha_k X_{ik} + e_{ij} \tag{8.10}$$

8.5.1.3 Random Model

The last model is to assess the significance of the constant and attribute models explained above. The model assumes that the image evaluation consists of the random component e_{ij} alone, that is,

$$U_{ij} = e_{ij} \tag{8.11}$$

which implies that the evaluation is the same for all department stores on average.

For the attribute model of each image item, only a subset of the five department store variables (X_{ik}; $k = 1,\cdots,5$) is considered in the model estimation process. The subset is preselected so as to include those that are most likely to be related to a particular image item. The model estimation process consists of the four steps described below.

1. Estimate the random model via the maximum likelihood method and obtain the associated AIC.
2. Estimate the attribute model using only one of the department store variables in the preselected subset for an image item under consideration.

Create one model for each variable in the subset. Select a model that has the smallest AIC value among these single-variable attribute models. If the AIC value of the selected model is smaller than that of the random model estimated in Step 1 by 3 or more, accept the attribute model as better than the random model. This decision criterion is based on the fact that the difference in the explanatory powers of two models is given by the difference in the associated AIC values, and the AIC difference that is greater than or equal to 2 or 3 is considered to be significant (Sakamoto et al., 1983). That is, when comparing models A and B, one can conclude that the explanatory power of model A is significantly higher than that of model B if the AIC value of model A is smaller than that of model B by 2 or 3.

3. Using the single-variable model selected in Step 2, estimate two-variable attribute models by adding one variable from the remaining subset of the department store variables. Select a model that has the smallest AIC value among the estimated two-variable models. If the selected model shows improvement of AIC by 3 or more, accept the selected model as better than the selected one-variable model.
4. Repeat Step 3 until AIC cannot be improved any more by using the remaining subset of the department store variables.

Variables whose parameters fluctuate considerably during this estimation process are not included in the final models so that their instability does not affect the models' reliability.

8.5.2 Results and Discussion

Results of the modeling are summarized in Table 8.5. The third and fourth columns show improvement in AIC for the constant and attribute models, respectively, in comparison with the random model. A large value in these columns indicates that the constant or attribute model can explain the common element of the department store image evaluation relatively well when compared with the random model, which assumes no influence of established evaluation specific to individual department stores or their measurable attributes. The rest of the table shows modeling parameters α_k (k = 1, …, 5) for the department store variables that are included in the final attribute model. It should be noted that most of the parameters are significant at either the 5% or 10% level, while the parameter selection has been based on AIC that represents overall explanatory power of the models.

Let us first look at the constant model, in which the element of the department store image evaluation that is common to all consumers is composed of a set of dummy variables representing individual department stores. This implies that image items that are successfully explained by the constant model have evaluation that has been well established with small variations between individual consumers. In Table 8.5, "luxuriousness," "elegant atmosphere," and "youthfulness/freshness" have the largest AIC improvement, and thus individuals' evaluation in terms of these image items is considered to have relatively good agreement between individuals. On the other hand, the smallest AIC improvement is found for "user-friendliness"

TABLE 8.5

Constant and Attribute Models: Improvement in AIC in Comparison with the Random Models and Department Store Variables Included in the Final Attribute Models

		Constant Model	Attribute Model					
				Attribute Parameters (t-Value)				
Category	Image Item	Improvement in AIC (χ^2 Value)	Improvement in AIC (χ^2 Value)	Square Footage X_{i1}	Designer-Brand Tenant X_{i2}	Years from Renovation X_{i3}	Nonsales Events X_{i4}	Years from Opening X_{i5}
Quality	Luxuriousness	518 (533.3 *)	215 (216.3 *)					0.0294 (14.608 *)
	Breadth of merchandise selection	210 (226.3 *)	72 (75.4 *)	0.1112 (7.249 *)		–0.0681 (–3.946 **)		
	Service quality of salespersons	332 (347.6 *)	196 (199.6 *)			–0.0458 (–2.586)		0.0267 (13.330 *)
	Elegant atmosphere	508 (524.7 *)	227 (229.3 *)					0.0299 (15.041 *)
	Attractiveness of nonsales events	221 (236.9 *)	190 (196.0 *)	0.0506 (3.007**)			0.0412 (8.090 *)	0.0155 (7.704 *)
Fashion	Fashionableness/urbaneness	167 (183.2 *)	62 (66.5 *)		–0.0372 (–3.196 **)	–0.1261 (–7.013 *)		
	Sensitivity to trends	236 (251.5 *)	69 (74.4 *)		–0.0339 (–2.670 **)	–0.1167 (–6.567 *)		–0.0072 (–3.865 *)
	Youthfulness/freshness	361 (376.9 *)	125 (129.3 *)			–0.1181 (–6.776 *)		–0.0183 (–10.111 *)
Usability	User-friendliness	113 (129.0 *)	55 (60.6 *)			–0.0627 (–3.756 *)	0.0265 (5.528 *)	–0.0124 (–6.255 *)
	Comfortable atmosphere for shopping	94 (109.6 *)	29 (30.8 *)			–0.0929 (–5.417)		

* = 5% significance level; ** = 10% significance level.

and "comfortable atmosphere for shopping," both of which belong to the usability category, indicating large variations in individual consumers' evaluation for these image items. These results conform to the findings about regional variations in the rankings of the department stores discussed in Section 8.3, where the former three image items have been found to have small variations and the latter two items to have large variations in the rankings by consumers utilizing different terminal stations. Moreover, also in Section 8.3, it has been pointed out that the image items in the usability category are likely to have their evaluation affected by relationships between individuals and department stores. This expectation corresponds to the findings here, too.

Turning now to the attribute model, we can see that evaluation for the quality image items, except for "breadth of merchandise selection," is relatively well explained by the five attributes of department stores used in this study. In particular, the "years since the store opened" variable not only has been included in four of the five models, but also has large *t*-values, which suggests strong influence of established tradition on department stores' image evaluation. This supports the finding in Section 8.3 that established tradition of department stores impacts on their evaluation. For the usability image items, on the other hand, the attribute model does not result in considerable improvement in AIC, so that the five attribute variables used in this study seem insufficient for explaining the evaluation associated with these items, and other attributes need to be considered. The fashion image items show intermediate relationships with the attribute variables of department stores, especially with the "years since the store opened" variable and the "years from the recent renovation" variable.

Comparison between the constant and attribute models shows that the constant model generally has better explanatory power than the attribute model for each image item regardless of its category. Difference in AIC improvement is rather large for the quality items, except for "attractiveness of the nonsales events," and the fashion items, especially "luxuriousness," "elegant atmosphere," and "youthfulness/freshness." The finding suggests that, for these image items, the common element of the department store image evaluation has a relatively small part explained by the stores' attributes. On the other hand, for the usability items and "attractiveness of the nonsales events," the small difference in AIC improvement can be interpreted as the common element that is largely explained by the attributes.

8.6 RELATIONSHIPS BETWEEN FREQUENCY OF VISITS AND THE IMAGE EVALUATION

In this section, we turn our attention toward the second element of the department store image evaluation models that varies between individual consumers. This element of the evaluation models is hypothesized to be determined by various relationships between department stores and individual consumers, such as accessibility, familiarity, store location, and consumer's residence. It is, however, very difficult to capture all such relationships in a measurable and thus usable manner for modeling, so we use the frequency of visits to department store i by individual j (Z_{ij2}) as a representative measure of the relationships. It is further assumed that department stores that offer convenience to individuals, and thus have good relationships with them, are more frequently visited, resulting in improved evaluation of the stores by the individuals. There are several reasons for this choice of the frequency of visits as the representative measure. First, the analysis in Section 8.3 has indicated that individuals tend to evaluate department stores located along the railway line they use or at its terminal station more highly than others, and this tendency appears to be better understood in relation to their use of those department stores than directly by the locational relationship between individuals and department stores. Second, other relationships that potentially influence the image evaluation, such as accessibility, familiarity, and usability, are likely to be integrated into the frequency variable. Third, preliminary analyses have indicated that this variable has better explanatory power than other variables, including the "travel time by rail" variable (Z_{ij1}).

We here develop models to explain the entire mechanism of the department store image evaluation, including both common and individually varying elements, by adding the relationship variable to the constant models that have been confirmed to have good explanatory power in the previous section. These models formulate the systematic component V_{ij} of the evaluation with the department store dummy variables ($X_{i6} \sim X_{i13}$) and the "frequency of visits" variable (Z_{ij2}). The following five models are created using different mathematical relationships between the frequency variable and the image evaluation.

Linear model, which assumes that the frequency of visits linearly proportionally impacts on the department store image evaluation:

$$U_{ij} = \sum_{k=6}^{13} \alpha_k X_{ik} + \gamma_2 Z_{ij2} + e_{ij} \tag{8.12}$$

where α_k ($k = 6, \ldots, 13$) and γ_2 are model parameters to be estimated.

Log-linear model, which assumes that the impact of the frequency of visits levels off as it takes a larger value:

$$U_{ij} = \sum_{k=6}^{13} \alpha_k X_{ik} + \gamma_2 \ln\left(Z_{ij2} + 1\right) + e_{ij} \tag{8.13}$$

Binary model, postulating that whether or not an individual visits a store in a given month strongly impacts on the evaluation, regardless of the frequency:

$$U_{ij} = \sum_{k=6}^{13} \alpha_k X_{ik} + \gamma_2 Z_{ij2} + e_{ij} \tag{8.14}$$

where $Z_{ij2} = 0$ if $Z_{ij2} = 0$, $Z_{ij2} = 1$ otherwise.

Store-specific log-linear model, where the positive impact of the frequency of visits is assumed to differ between department stores:

$$U_{ij} = \sum_{k=6}^{13} \alpha_k X_{ik} + \gamma_{i2} \ln\left(Z_{ij2} + 1\right) + e_{ij} \tag{8.15}$$

where γ_{i2} ($i = 1, \ldots, 9$) are store-specific parameters.

Store-specific binary model, where the positive impact of visits (regardless of the frequency) is assumed to differ between department stores:

$$U_{ij} = \sum_{k=6}^{13} \alpha_k X_{ik} + \gamma_{i2} Z_{ij2} + e_{ij} \tag{8.16}$$

Table 8.6 summarizes improvement in AIC by the incorporation of the "frequency of visits" variable in comparison with the constant model. The AIC values have been significantly improved in the majority of the estimated models, indicating that the individually varying element of the department store image evaluation is closely associated with individuals' use of the department stores.

TABLE 8.6
Improvement in AIC in Comparison with the Constant Models

Category	Image Item	Linear	Log-Linear	Binary	Store-Specific Log-Linear	Store-Specific Binary
Quality	Luxuriousness	11	36	43	35	40
	Breadth of merchandise selection	12	28	27	26	25
	Service quality of salespersons	2	16	19	16	8
	Elegant atmosphere	2	8	9	9	10
	Attractiveness of nonsales events	13	32	33	39	31
Fashion	Fashionableness/urbaneness	6	19	22	13	14
	Sensitivity to trends	7	16	17	7	10
	Youthfulness/freshness	4	5	4	2	–6
Usability	User-friendliness	90	119	85	133	110
	Comfortable atmosphere for shopping	38	83	81	82	85

Note: Shaded cells indicate a model with the best AIC value for each image item.

By comparing the first three models that do not incorporate store-specific parameters, we can see that the log-linear and binary models have much better explanatory power than the linear model for most image items. This result supports the hypotheses incorporated in the log-linear and binary models; that is, the relationship between the department store image evaluation and the frequency of visits is not linear, and the impact of the frequency of visits on the evaluation levels off as the frequency increases (the log-linear model) or depends only on whether or not an individual visits a store (the binary model). Of the four models reflecting these two hypotheses, however, the best model (i.e., the model showing the largest improvement in AIC) differs from image item to image item, so that it cannot be determined which hypothesis is more appropriate to explain the relationship between the evaluation and the frequency of visits. In addition, the log-linear and binary models show larger AIC improvement for slightly more image items than the store-specific log-linear and binary models do, which suggests the relationship does not vary considerably between the nine department stores examined in this study.

In these department store image evaluation models, it is assumed that the "frequency of visits" variable (Z_{ij2}) explains the evaluation element that varies between individuals, while the department store variables ($X_{i6} \sim X_{i13}$) explain the other evaluation element that is common to all consumers. Accordingly, the improvement in AIC obtained by the inclusion of the "frequency of visits" variable can be interpreted as the proportion that the individually varying element accounts for in the department store image evaluation. The usability image items of "user-friendliness" and "comfortable atmosphere for shopping," which indicate especially large improvement in AIC, can therefore be considered to have image evaluation that fluctuates greatly between individual consumers, depending on their use of each department store. Some potential reasons for this tendency include, but are not limited to, the fact that consumers may naturally feel some sense of closeness toward a department store they often visit, and that frequent visits to a particular store familiarize themselves with it, which may well result in increased usability of the store. On the other hand, "youthfulness/freshness" and "elegant atmosphere" indicate very minor improvement in AIC, so that their evaluation appears mostly constant among individual consumers with little variation. These results correspond well with those in Sections 8.3 and 8.5.

Let us next examine how the variation in the image evaluation between individual consumers differs with department stores. The magnitude of the variation is represented by the store-specific parameters for the "frequency of visits" variable (γ_{i2}; $i = 1, \ldots, 9$) in the store-specific log-linear and binary models. Because there is no qualitative difference between the two models with respect to their explanatory power, the discussion here only considers the store-specific binary model, focusing on whether or not an individual has ever visited a particular department store in a given month.

The estimated model parameters are listed in Table 8.7. First, we can see that "luxuriousness," "user-friendliness," and "comfortable atmosphere for shopping" tend to have large absolute values for the store-specific parameters, implying large variation in individuals' image evaluation for these image items. This corresponds with the previous results very well as regards the second and third image items. For these two image items, Odakyu in Shinjuku, Tobu in Ikebukuro, and Takashimaya in Futakotamagawa are given large positive parameter values, which indicates the same tendency found in Section 8.3; that is, the image evaluation tends to be better for department stores that are located along the railway line that individuals use or at its terminal station. On the other hand, "youthfulness/freshness" and "elegant atmosphere" have relatively small parameter values, again corresponding well with the results discussed so far. In terms of the department stores, Tobu in Ikebukuro is given particularly large parameter values for most image items. This may be explained by the fact that it is one of the major department stores in the Ikebukuro commercial area, where the terminal station of the Tobu-tojo line is located. In addition, the common element of the image evaluation associated with this store might have been underestimated because a relatively small proportion of the survey respondents reside along the Tobu-tojo line.

8.7 CONCLUSIONS AND FUTURE RESEARCH QUESTIONS

This study analyzed consumers' evaluation of department stores in terms of selected image items on the basis of the ranking data obtained by the questionnaire survey. The major findings of this study are summarized below.

1. The image items considered in this study can be divided into those for which evaluation is relatively constant among individual consumers and those for which evaluation varies greatly between individuals, and the variation in the latter appears attributable to individuals' use of each department store.
2. The department stores considered in this study can also be divided into the same two groups. Department stores for which evaluation considerably varies between individuals tend to be located at the terminal stations or along the railway lines in the study area of the questionnaire survey, while those for which evaluation is relatively fixed tend to have a large market area and generally receive rather high evaluations.
3. The image evaluation of department stores can be decomposed into an element that is common to all consumers and another element that varies between consumers, and which element has stronger influence differs from image item to image item.
4. The common element in the department store image evaluation for certain image items can be explained by characteristics of department stores, such as years since the store opened.
5. The element that varies between individual consumers can be assumed to be explained by general relationships between individuals and department stores, and the frequency of visits to each department store appears to be a relatively good proxy of such relationships. While the evaluation tends to improve as the frequency of visits increases, the relationship is not linear. The influence of the frequency of visits can be assumed either to level off or to depend only on whether an individual has visited a particular department store at least once in a given month.

TABLE 8.7
Estimated Parameters for the "Frequency of Visits" Variable in the Store-Specific Binary Models

Category	Image Item	Takashimaya (N)	Mitsukoshi (N)	Matsuzakaya (G)	Isetan (Sj)	Odakyu (Sj)	Seibu (Sh)	Tokyu (Sh)	Tobu (I)	Takashimaya (F)
Quality	Luxuriousness	1.447	0.106	0.724	0.618	0.482	0.613	0.689	0.833	1.205
	Breadth of merchandise selection	0.869	0.156	0.345	0.367	0.452	0.435	0.440	1.600	0.653
	Service quality of salespersons	0.545	0.712	0.967	0.400	0.304	0.376	0.275	0.936	0.337
	Elegant atmosphere	0.581	–0.428	–0.547	0.357	0.223	0.175	0.689	0.923	0.549
	Attractiveness of nonsales events	0.916	0.823	0.656	0.535	0.262	0.064	0.513	1.253	0.825
Fashion	Fashionableness/urbaneness	0.741	0.405	–0.755	0.617	0.415	0.559	0.420	0.415	0.713
	Sensitivity to trends	0.722	0.200	–0.244	0.034	0.564	0.486	0.298	0.651	0.716
	Youthfulness/freshness	0.255	0.298	–0.177	0.575	–0.073	0.236	0.034	0.633	0.293
Usability	User-friendliness	0.908	0.705	0.437	0.946	1.353	0.426	0.287	2.152	1.358
	Comfortable atmosphere for shopping	0.758	0.807	0.498	0.819	1.182	0.397	0.650	1.681	1.590

Note: Shaded cells indicate parameter values greater than 1.

In this study, we could not fully investigate the structure of the evaluation element common to all consumers. There are many attributes of department stores that potentially influence them other than those considered here, for example, sales volume, the number of employees, the number of stores owned by the same company, advertising costs, square footage, and sales volume per merchandise category, to name but a few. It is thus necessary to examine other potential variables to better understand the common element in the department store image evaluation. Further, improvement of the evaluation models may also be attained by considering other expressions that would better reflect the influence of each attribute than the linear expression currently in use.

We believe that the findings obtained in this study can be applied to creating images of department stores that would resolve or ease the existing controversy between them and other large-scale retailers. The methodology used is also suitable for examining how other factors, such as availability of parking facilities and accessibility by public and private transportation, affect consumers' evaluation of commercial facilities. Further investigation of various aspects of consumers' image of commercial facilities is awaited so that issues associated with modern shopping environments can be understood in a more comprehensive manner.

ACKNOWLEDGMENTS

We wish to express our sincere thanks to Dr. Atsuyuki Okabe for supporting us with inspiring conversations and valuable comments. This research was not possible without the kind support of the survey respondents; we appreciate the effort and time they devoted to completing the questionnaire.

This chapter was first published in the *Journal of Applied Regional Science* in Japanese and translated into English for republication in this volume by kind permission of the Applied Regional Science Conference, the publisher of the journal. Minor modifications have also been made for the purpose of clarification.

Original publication: Yamada, I., and Sadahiro, Y. 1998 "An Empirical Analysis of Consumers' Evaluation on Department Stores," *Journal of Applied Regional Science* 3: 49–60.

REFERENCES

Asada, H., Nagai, K., and Kon'no, A. 1991. The use of Toyokawa's central shopping district and environs: Observations on users and customer's behavioral patterns. *Papers of the Annual Conference of the City Planning Institute of Japan* 26:889–94 (in Japanese).

Bearden, W. O. 1977. Determinant attributes of store patronage: Downtown versus outlying shopping centers. *Journal of Retailing* 53:15–22.

Bellenger, D. N., Robertson, D. H., and Greenberg, B. A. 1977. Shopping center patronage motives. *Journal of Retailing* 53:29–38.

Ben-Akiva, M., and Lerman, S. R. 1985. *Discrete choice analysis: Theory and application to travel demand.* Cambridge, MA: MIT Press.

Burt, J. E., and Barber, G.M. 1996. *Elementary statistics for geographers.* 2nd ed. New York: Guilford Publications.

Copeland, M. T. 1923. The relation of consumers' buying habits to marketing methods. *Harvard Business Review* 1:282–89.

Domencich, T. A., and McFadden, D. 1975. *Urban travel demand: A behavioural analysis*. Amsterdam: North-Holland Publishing.

Fararo, T. J. 1978. *Mathematical sociology: An introduction to fundamentals*. Huntington, NY: R.E. Krieger.

Heidingsfield, M. S. 1949. Why do people shop in downtown department stores? *Journal of Marketing* 13:510–12.

Hirschman, E. C. 1979. Intratype competition among department stores. *Journal of Retailing* 55:20–34.

Honda, H. 1983. Basic research for the analysis of preference for shopping areas. *Papers of the Annual Conference of the City Planning Institute of Japan* 18:463–67 (in Japanese).

Huff, D. L. 1963. A probabilistic analysis of shopping center trade area. *Land Economics* 39:81–90.

Japan Society of Traffic Engineers. 1993. *Disaggregate travel demand analysis made simple*. Tokyo: Japan Society of Traffic Engineers (in Japanese).

Koide, O. 1976. Spatial movement model and consumer behavior theory. *Papers of the Annual Conference of the City Planning Institute of Japan* 11:289–94 (in Japanese).

Kondo, A., and Aoyama, Y. 1989. A comparative study on consumer shopping behavior to shopping centers in city center and suburb and an analysis of demand for them. *Papers of the Annual Conference of the City Planning Institute of Japan* 24:565–70 (in Japanese).

Luce, R. D. 1959. *Individual choice behavior: A theoretical analysis*. New York: Wiley.

Morichi, S., Yai, T., Fujii, T., and Takeuchi, K. 1984. Analysis of destination choice behavior for non-grocery shopping. *Infrastructure Planning Review* 1:27–34 (in Japanese).

Nishisato, S. 1982. *Quantification of qualitative data*. Tokyo: Asakura Shoten.

Rich, S. U. 1963. *Shopping behavior of department store customers*. Boston: Division of Research, Graduate School of Business Administration, Harvard University.

Sakamoto, Y., Ishiguro, M., Kitagawa, G., and Kitagawa, T. 1983. *Information criterion statistics*. Tokyo: Kyoritsu Shuppan (in Japanese).

Sasaki, Y., Omi, T., Yamada, H., and Sugai, T. 1980. A study of variations in shopping districts from the standpoint of behavior choosing purchasing zones. Part I. The model of consumer's purchasing behavior. *Transactions of the Architectural Institute of Japan* 291:71–78 (in Japanese).

Stewart, J. Q. 1948. Demographic gravitation: Evidence and applications. *Sociometry* 11:31–58.

Sugioka, S. 1991. *Law on large-scale retail facilities, urban commerce, and citizens: An introduction to commercial accumulation policy*. Tokyo: Nippon Hyoronsha (in Japanese).

Takaoka, S., and Koyama, S. 1991. *Modern department stores*. A revised edition. Tokyo: Nihon Keizai Shinbunsha (in Japanese).

Upton, G. J. G. 1978. *The analysis of cross-tabulated data*. Probability and Statistics Series. Chichester, UK: John Wiley.

Weisbrod, G. E., Parcells, R. O., and Kern, C. 1984. A disaggregate model for predicting shopping area market attraction. *Journal of Retailing* 60:65–83.

9 An Experimental Analysis of the Perception of the Area of an Open Space Using 3-D Stereo Dynamic Graphics

Toru Ishikawa, Atsuyuki Okabe, Yukio Sadahiro, and Shigeru Kakumoto

9.1 INTRODUCTION

In a densely populated city such as Tokyo, open space is very precious; people feel as if open space is an oasis in a crowded, built-up environment. With no budget constraints, it is thus desirable for people to have a large open space. In practice, however, land price in a densely populated city is so high that it is often difficult to have a large open space. When the physical area of an open space is fixed, maximizing the perceived area of the space is one of the feasible ways to provide people with a feeling of comfort or spaciousness. It is hypothesized that the perceived area of an open space would vary according to the shape and location of the space even if its physical area remains the same. The objective of this chapter is to find the best shape/location of an open space, that is, the shape/location that gives the largest perceived area.

This chapter experimentally investigates the perception of the area of an open space in a city using 3-D stereo dynamic computer graphics. With regard to the perception of area, many experimental studies have been conducted, especially focusing on the relationship between shape and perceived area. For example, Anastasi (1936) found that the perception of the area of geometric figures having the same area varies according to their shapes. He found that the perceived area was positively correlated with both the perimeter and the diameter. Holmberg and Holmberg (1969) found that as the height/width ratio of a rectangle increased, its perceived size increased. On the other hand, Holmberg and Wahlin (1969) found no relationship between the shape and perceived size of a rectangle. In addition, Smith (1969) reported that the perceived area of a figure was negatively correlated with its perimeter. Martinez and Dawson (1973) further examined the effect of shape on the perception of area. They classified the shape of figures into six categories (i.e., triangles, quadrilaterals, polygons, ellipses, stars, and irregular figures) and examined the relationship between the shape/perimeter and perceived area within/between categories. Within all categories except triangles and irregular figures, the perceived area was negatively correlated with the perimeter. In contrast, when figures belonged to different categories, no relationship was found between the perceived area and the perimeter.

The above studies dealt with the shape of figures drawn on or made of a piece of paper. The relationship between perceived area and shape at the scale of a building was explored by Sadalla and Oxley (1984). They investigated the perception of the area of a room. In their study, subjects were shown rooms of the same area having different ratios of length to width and were asked to estimate the area of the rooms. The results indicated that the area of a room was perceived as larger as the length/width ratio increased. This finding is consistent with the results of some of the aforementioned studies using figures drawn on a piece of paper (Anastasi, 1936; Holmberg and Holmberg, 1969; Martinez and Dawson, 1973).

Benedikt and Burnham (1985) explored the perception of area in a more systematic way. They termed a space visible from a specific vantage point the *isovist* and developed five measures to characterize 2-D isovists. On a laboratory table, they constructed 560 spaces of different shapes with the same total area (4.5 square feet) but different isovists by varying the values of isovist measures. Subjects saw a pair of spaces and judged which space had more visible area and which space had a larger total area taking invisible area into account. The area of isovist (i.e., visible area) was found to have a significant effect on the estimation in both cases. The length of the visible boundary of the isovist also affected the estimation.

The present study is concerned with the perception of area at a city scale. Few studies have been conducted at a city scale, however. To fill this gap in research, this study aims to investigate the perception of the area of an open space at a city scale. As mentioned above, this study is of much importance for the planning of a spacious city under a limited budget. Understanding the perception of area makes it possible to provide more spaciousness using an open space with a constant area. First, the chapter experimentally investigates the perception of the area of an open space in a city with respect to its shape and location. Second, the chapter formulates three possible models for explaining the perception of area. Last, the chapter statistically tests those models and finds the best model.

Three-dimensional stereo dynamic computer graphics (3-D stereo geographic information systems [GIS]) were used to conduct this study. This device has the following advantages:

1. It controls possible factors affecting the perception of the area of an open space.
2. It creates a hypothetical reality of walking in an actual city.
3. It enables one to collect data in a less costly way than in a field setting.

The first advantage played an important role in this experiment. In an actual space, many factors may affect the perception of the area of an open space, and controlling these factors is almost impossible. Moreover, it is very costly to construct a city-scale experimental field. To overcome the difficulties encountered in actual field experiments, some experiments of spatial cognition and perception have recently been conducted with a computer-simulated environment (e.g., Golledge et al., 1995; O'Neill, 1992; Tlauka and Wilson, 1994, 1996). These studies constructed a city using computer graphics and presented a simulated view to subjects. Computer graphics made it possible to construct a city and control variables to fulfill the purpose of the experiment.

In addition to the advantage of controllability, the device used here presents a 3-D view (when seen through polarizing glasses) and has a walk-through function. These advantages make it possible to conduct an experiment easily and to make subjects feel as if they were walking in an actual city. In the present study, the objective is to investigate the perception of the area of an open space in a city—to conduct an experiment making subjects feel as if they were in an actual city. Thus, it is important to present subjects with a 3-D view and have subjects walk in a simulated city. The 3-D view and the walk-through function are major characteristics of the device, which cannot be realized using static computer graphics.

9.2 METHOD

The experiment was conducted on July 22, 25, 26, 27, and 28, 1994. The experiment consisted of 57 participants (34 males and 23 females); their ages ranged from 19 to 59 years. On a computer screen, buildings were projected as shown in Figure 9.1. The subjects watched the screen wearing polarizing glasses. Through polarizing glasses, slightly different views are shown to each eye of the subjects, thus creating a hypothetical reality of being in an actual 3-D space. Each subject was asked to walk through a simulated city consisting of two blocks (a plan is shown in Figure 9.2). The

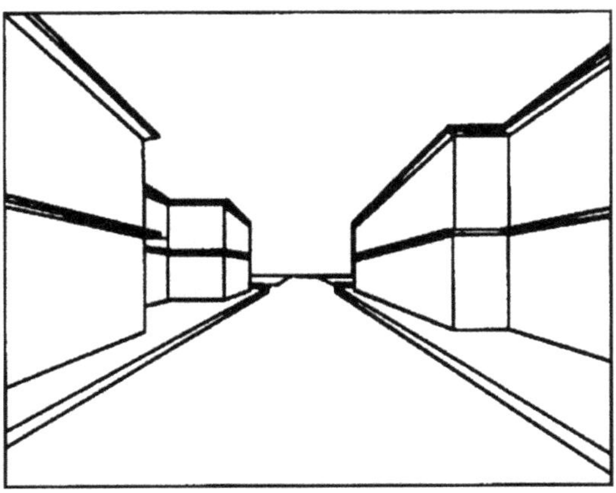

FIGURE 9.1 Perspective view of a simulated environment. Subjects saw this on a screen wearing polarizing glasses, so they saw it in a 3-D perspective. On the screen, it was shown in color, with the buildings dark yellow, the blocks light gray, the streets dark gray, and the sky light blue.

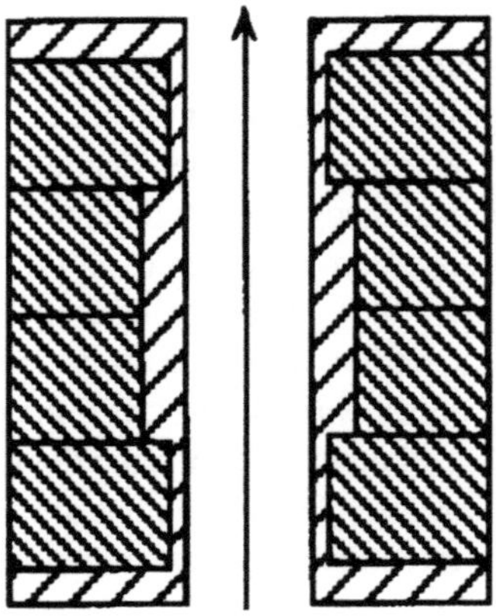

FIGURE 9.2 Plan of blocks. The heavily hatched area is covered with buildings; the lightly hatched area is open space. The arrow shows a walking direction. Subjects did not see this plan; they saw the perspective view in Figure 9.1.

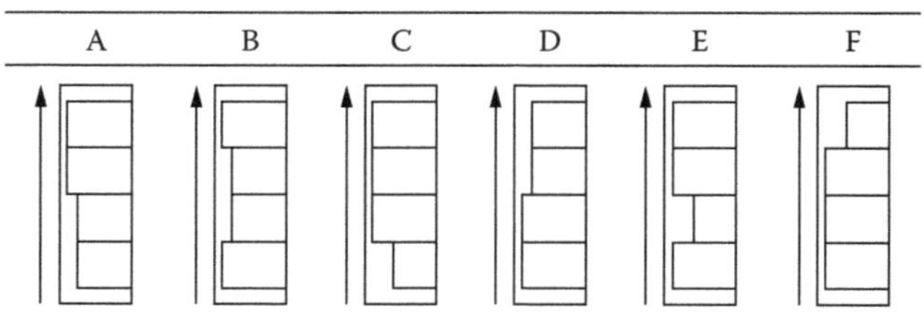

FIGURE 9.3 Six patterns of blocks (A–F) with respect to the shape and location of an open space. The arrows shows walking directions.

heavily hatched regions are occupied by two-story buildings and the lightly hatched regions indicate open space (not occupied by buildings). The perspective view of the buildings is shown in Figure 9.1. The subjects saw this 3-D view (not a plan as shown in Figure 9.2) during the walk.

Two parameter values were fixed throughout the experiment. First, the area of a region occupied by buildings on a block is the same for all blocks (i.e., the heavily hatched area on the left-hand side block is the same as that on the right-hand side block in Figure 9.2). Second, the area of open space is the same for all blocks (i.e., the lightly hatched area on the left-hand side block is the same as that on the right-hand side block in Figure 9.2).

Control variables were the shape and location of an open space. Six patterns of blocks with respect to the shape and location of an open space (Patterns A to F in Figure 9.3) were used. Using these patterns, 6 pairs of blocks (consisting of 12 blocks in total, Blocks B1 to B12) were constructed and 12 walks for each subject (Walks 1 to 12) were conducted, as shown in Figure 9.4. The order of the 12 walks was counterbalanced across subjects to avoid an order effect.

Before the experiment, instructions were given to the subjects so that they understood the open-space ratio, which is defined by the area of open space divided by

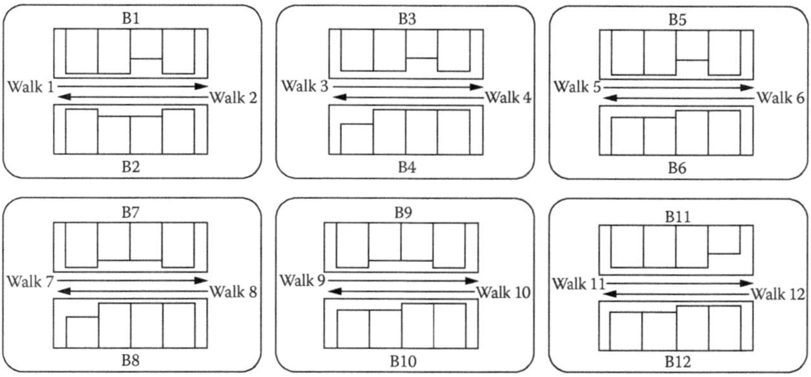

FIGURE 9.4 Six pairs of blocks consisting of twelve blocks (B1–B12). The arrows indicate twelve walks (Walks 1–12). Each subject conducted one walk at a time.

the area of a block. Thus, subjects understood the meaning of the open-space ratio completely. Each subject watched the simulated views, walking from one end of the street to the other end, and was then asked to answer the following two questions:

Question 1. Which side has more open space: the right-hand side, the left-hand side, or do they both have the same amount?
Question 2. For each block, please estimate the open-space ratio.

Each subject answered the above two questions on an A4-size sheet of paper given out beforehand. The experiment took each subject approximately 25 minutes.

9.3 RESULTS

9.3.1 Paired Comparison of the Area of an Open Space between the Right-Hand Side and Left-Hand Side

Answers to Question 1 are presented in Table 9.1. The first row shows the number of subjects who answered that the right-hand (R) side had more open space. The second row shows the number of subjects who answered that the left-hand (L) side had more open space. The third row shows the total number of subjects who answered the R side or the L side. The fourth row shows the number of subjects who answered that the R side and L side had the same area of open space. The difference of the perception between the R side and L side was statistically tested using the binomial distribution. In the following examinations in this section and the next, a space created by a setback is referred to as an open space (enclosed by the heavy lines in Figure 9.5).

Let us examine the implications of Table 9.1 with respect to the shape and location of an open space (see Figure 9.4). First, Walks 1 and 2 were inspected. Both Blocks B1 and B2 have an open space at the center, but the shapes of the open spaces are different from each other. Block B1 has an open space with a deep setback and a narrow frontage, whereas Block B2 has an open space with a shallow setback and a wide frontage. The first and second columns of Table 9.1 show that the open space

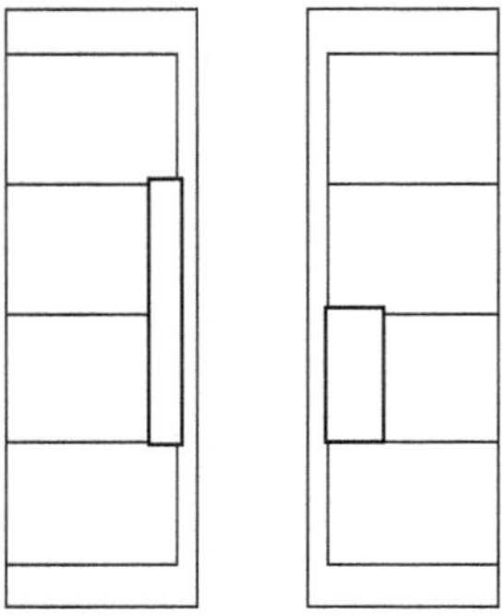

FIGURE 9.5 Spaces created by a setback (enclosed by the heavy lines).

TABLE 9.1
Results of the Paired Comparisons

	Walk Number											
Answers	**1**	**2**	**3**	**4**	**5**	**6**	**7**	**8**	**9**	**10**	**11**	**12**
Right[a]	42**	10**	34*	33**	41**	23	22	43**	34**	30**	39**	23
Left[b]	13**	31**	18*	7**	12**	21	21	10**	12**	6**	9**	20
Right or left[c]	55**	41**	52*	40**	53**	44	43	53**	46**	36**	48**	43
Same[d]	2**	16**	5*	17**	4**	13	14	4**	11**	21**	9**	14

[a] This row represents the number of subjects who answered that the right-hand side had more open space.
[b] This row represents the number of subjects who answered that the left-hand side had more open space.
[c] This row represents the total number of subjects who answered that the right-hand side or the left-hand side had more open space.
[d] This row represents the number of subjects who answered that both sides had the same amount of open space.
* = $p < .05$; ** = $p < .01$.

on Block B2 is perceived as larger than that on Block B1 in both Walks 1 and 2. This indicates that the shape of an open space has an effect on the perception of the area; specifically, an open space with a wide frontage is perceived as larger than that with a deep setback.

Second, Walks 3 and 4, and Walks 9 and 10 were inspected. The shapes of open spaces on both sides are the same, but the location on each block is different. Blocks B3 and B9 have an open space at the center of the blocks, whereas Blocks B4 and B10 have an open space at the end. (The pair of Walks 3 and 4 is compared in the third and fourth columns of Table 9.1.) In Walk 3, the open space on Block B4 (R side) is perceived as larger than that on Block B3 (L side). In Walk 4, the walking direction is reversed and the open space on Block B3 (R side) is perceived as larger. These results imply that the perception of the area of an open space is affected by the location of the space on a block. A comparison of the pair of Walks 9 and 10 reveals similar results (see the ninth and tenth columns of Table 9.1). In Walk 9, the open space on block B10 (R side)

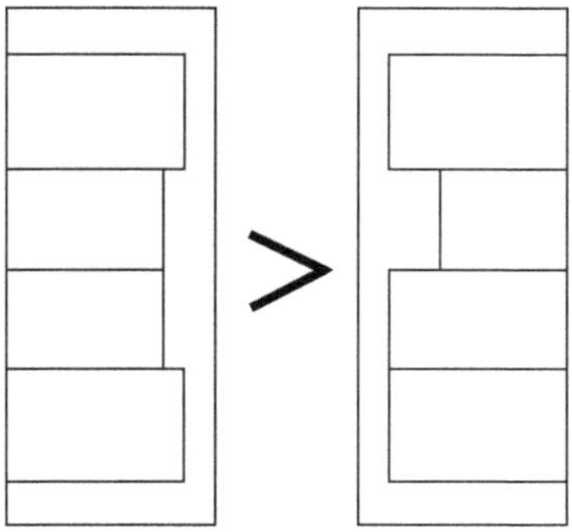

FIGURE 9.6 Width/depth ratio effect. An open space with a wide frontage (left) is perceived as larger than that with a deep setback (right).

is perceived as larger. In Walk 10, the walking direction is reversed and the open space on Block B9 (R side) is perceived as larger. These results show that it can be concluded that the location of an open space on a block affects the perception of the area.

Next, the effect of a location was examined in more detail. In Walks 3 and 9, the open spaces of the R side (Blocks B4 and B10) are located in such a way that subjects can see the whole area of the spaces. Thus, the area of the open space that subjects can see is larger on Blocks B4 and B10 than on Blocks B3 and B9. However, when the walking direction is reversed (in Walks 4 and 10), the open spaces on Blocks B4 and B10 are located at the end of the blocks. Located at the end of the blocks, the visible area of these open spaces becomes less than that of the L side (Blocks B3 and B9). This examination reveals that the effect of a location on perceived area results from the visible area of an open space. Specifically, as the visible area of an open space becomes larger, the perceived area of the open space increases.

The obtained results can be summarized as follows:

1. An open space with a wide frontage is perceived as larger than an open space with a deep setback, which will be referred to as the *width/depth ratio effect* (see Figure 9.6).
2. As the visible area of an open space becomes larger, the perceived area of the open space increases, which will be referred to as the *visible area effect* (see Figure 9.7).

9.3.2 Estimation of the Open-Space Ratios

Next, the answers to Question 2 were analyzed. To compare the estimated values of the open-space ratios with one another, each subject's estimated values for 12 blocks were standardized. Note that the estimated open-space ratios for the blocks of the same open-space pattern (e.g., estimated values for Blocks B2, B7, and B9 in Figure 9.4) were averaged.

In Table 9.2, blocks are ranked by the estimated open-space ratios. The order of the estimated open-space ratios can be explained clearly by focusing on the shape and location of an open space (for Patterns A to F, see Figure 9.3). An open space

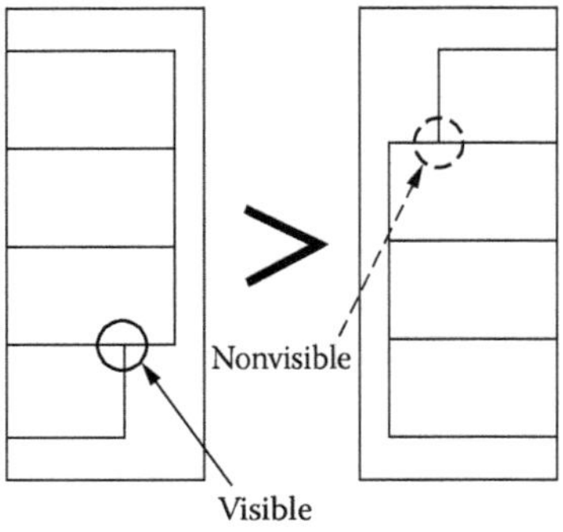

FIGURE 9.7 Visible area effect. As the visible area of an open space becomes larger, the perceived area increases.

TABLE 9.2
Estimated Value of the Open-Space Ratio for Each Pattern

	Pattern					
	A	B	C	D	E	F
Average	0.31	0.26	0.08	0.04	–0.25	–0.45

of Pattern A has a wide frontage and a shallow setback. This open space is located at the front of a block, and hence subjects can see the whole area of the open space. An open space of Pattern B has a wide frontage and a shallow setback. This open space is located in such a way that subjects can see its partial area. An open space of Pattern C has a narrow frontage and a deep setback. Subjects can see the whole area of this open space. An open space of Pattern D has a wide frontage and a shallow setback. This open space is located at the end of a block, and hence the visible area of the open space is very small. An open space of Pattern E has a narrow frontage and a deep setback. This open space is located in such a way that subjects can see its partial area. An open space of Pattern F has a narrow frontage and a deep setback. This open space is located at the end of a block, and hence the visible area of the open space is very small.

The above-mentioned findings can be summarized as follows:

1. An open space with a wide frontage is perceived as larger than an open space with a deep setback (width/depth ratio effect).
2. As the visible area of an open space becomes larger, the perceived area of the open space increases (visible area effect).

The results obtained in this section are consistent with those obtained in the previous section. It is important to note that the same results were obtained from two different kinds of questions. One (in the preceding section) is a task of comparing two open spaces and judging which open space is larger (paired comparison of area). The other (in this section) is a task of estimating the open-space ratios (numerical judgment of area).

9.3.3 Perception Models of the Area of an Open Space

Keeping the width/depth ratio effect and the visible area effect in mind, the attention was then focused on formulating models that explain the perception of the area of an open space. First, two factors regarding perception were discussed. When objects in space are observed with a person's head fixed facing straight ahead, the person does not perceive the area of 360 degrees; objects are perceived in a limited area called the *visual field*. The visual field is said to be the area of about 90 degrees to the right and left of the center. In the present experiment, the visual field of a subject was limited to the hatched area in Figure 9.8. Thus, the following discussion assumes that a subject perceives the area of 70 degrees to the right and left of the center.

In addition to the visual field factor, one more factor should be considered. When objects in space are observed, they are perceived in a perspective view. To deal with this view mathematically, a transformation called the *viewing transformation* or the *perspective view transformation* is used. The viewing transformation assumes that a viewing point is at the origin and that there is a plane at the focal distance (called the *image plane*). The image plane corresponds to the retina or camera film. Images of objects observed in a perspective view can be obtained by projecting the objects onto the image plane (see Figure 9.9, in which the focal distance is 1).

A subject perceives area continuously during the walk, and hence the amount of perceived area varies as a subject walks. With regard to the perception of area, three possible models were considered:

1. A subject perceives all the visible area.
2. A subject perceives an area within a limited range.
3. A subject perceives the largest visible area of a setback space.

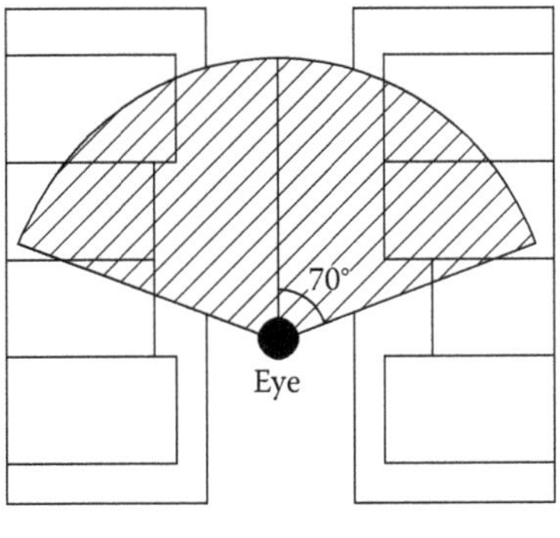

FIGURE 9.8 Visual field. The subjects perceived the area of 70 degrees to the right and left of the center (hatched area).

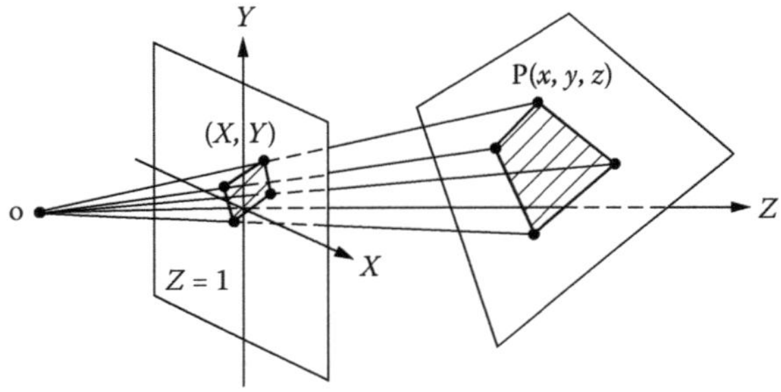

FIGURE 9.9 Viewing transformation. P(x,y,z) is projected on the image plane at the focal distance, which is 1 here ($X = x/z$, $Y = y/z$).

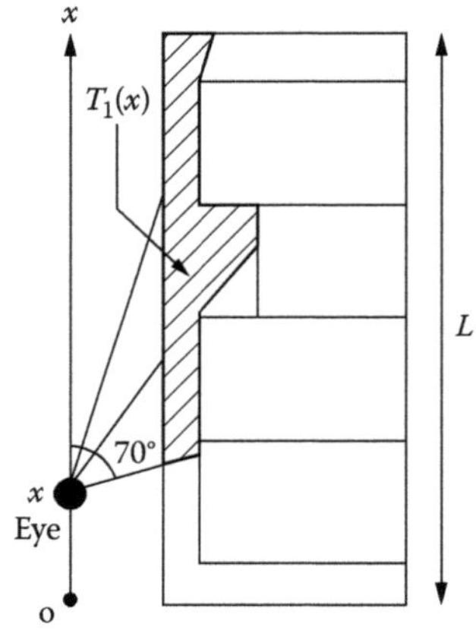

FIGURE 9.10 All the visible area at a point x (seen from above).

Once a model is chosen, an explicit form of the perceived area, $S(x)$, is derived at a point x. If an explicit form of $S(x)$ is given, then the total amount of perceived area, called the *total perceived area* (S^*), during the walk along L can be obtained from the following equation:

$$S^* = \int_0^L S(x)\,dx \tag{9.1}$$

The focus turned to deriving an explicit form of $S(x)$ for the above three models (Models 1, 2, and 3) and statistically testing which model showed a good fit to the observed data.

9.3.3.1 Model 1: A Subject Perceives All the Visible Area

This model assumes that a subject perceives all the visible area at every point along L. The visible area at a point x is indicated by the region enclosed by the heavy lines in Figure 9.10. This area is denoted by $T_1(x)$. It should be noted here that Figure 9.10 is a plan seen from above. The perceived area, $S_1(x)$, should be considered in a perspective view. Thus, $T_1(x)$ should be transformed into $S_1(x)$ through the viewing transformation (F), that is,

$$S_1(x) = F(T_1[x]) \tag{9.2}$$

Having obtained the perceived area at a point x from Equation 9.2, the total perceived area can be obtained from Equation 9.1, in which $S_1(x)$ is substituted for $S(x)$. In this manner, the total perceived area can be obtained for Patterns A to F, respectively (Figure 9.3). From the experiment, as shown in Table 9.2, the estimated values of the open-space ratios for those patterns were obtained. Using these two sets of values, the correlation coefficient was computed between the two variables (.42). This is not significant at the .05 level.

9.3.3.2 Model 2: A Subject Perceives an Area within a Limited Range

Model 1's assumption that a subject perceives all the visible area even if the area is located far away is somewhat unrealistic, and this may have caused its nonsignificance. The range of a person's view must be limited; that is, a person must perceive objects within a limited range ahead of a standing position. This range is called the *perception range*. The proposed model next takes the perception range into account.

This model assumes that the perception range is within 10 m ahead of a viewing point (e.g., Higuchi, 1975, p. 47). The perceived area at a point x is indicated by the area enclosed by the heavy lines in Figure 9.11, and this area is denoted by $T_2(x)$. Note again that Figure 9.11 is a plan seen from above. The perceived area, $S_2(x)$, in a perspective view should be considered. Hence, $T_2(x)$ is transformed into $S_2(x)$ through Equation 9.2, and the total perceived area, S^*, is obtained from Equation 9.1, in which $S_2(x)$ is substituted for $S(x)$. The total perceived area for Patterns A to F can be calculated in this manner, and Table 9.2 shows the estimated values of the open-space ratios for those patterns. The correlation coefficient between these two variables is –.16. This is also not significant at the .05 level.

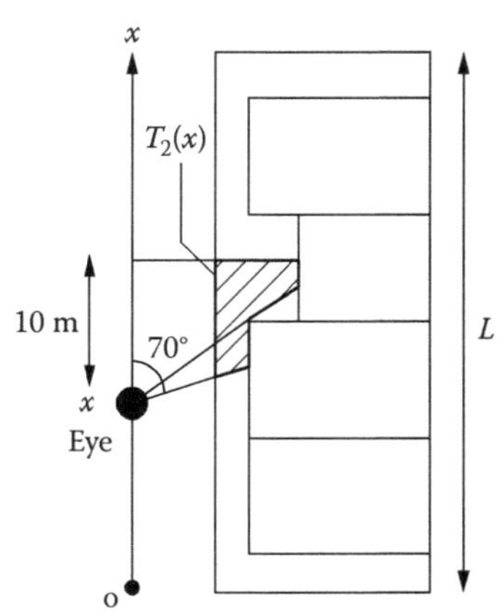

FIGURE 9.11 Area within the perception range (10 m).

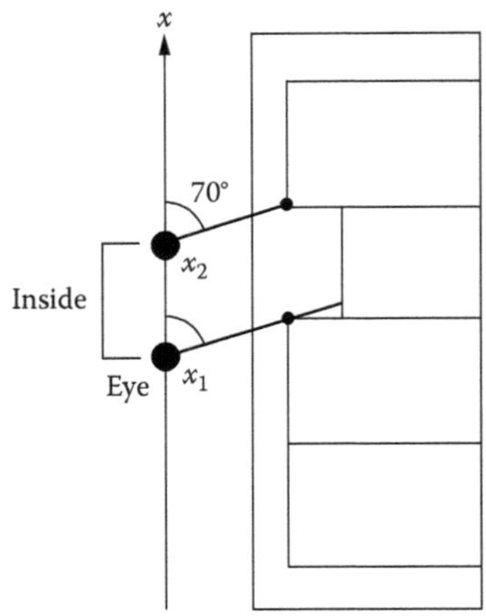

FIGURE 9.12 Inside/outside of a setback space. A subject considers himself or herself to be inside the setback space while he or she is between points x_1 and x_2.

9.3.3.3 Model 3: A Subject Perceives the Largest Visible Area of a Setback Space

The previous two results suggest that the perceived area cannot be obtained by merely integrating the visible area. Thus, a third perception model was developed. On each block of the simulated city, the wall lines of the buildings were not straight and there was a space created by a setback (enclosed by the heavy lines in Figure 9.5). In this model, this space is called a *setback space*, and attention is focused on it.

This leads to two new notions: inside and outside of a setback space (see Figure 9.12). When a subject reaches a point x_1, a setback space occupies the entire visual field. At a point x_1, a subject considers that he or she enters inside the setback space. During the walk between points x_1 and x_2, a subject considers that he or she is inside the setback space. When a subject reaches a point x_2, the setback space goes out of the visual field completely. At a point x_2, a subject considers that he or she goes out of the setback space; that is, he or she is outside the setback space.

At a point x_1, a subject perceives the largest visible area of the setback space (enclosed by the heavy lines in Figure 9.13). This area is denoted by T_3. Note that Figure 9.13 is a plan seen from above and that the perceived area, S_3, should be considered in a perspective view. Hence, S_3 is obtained from the following:

$$S_3 = F\left(T_3\right) \tag{9.3}$$

where F is the viewing transformation.

A subject regards S_3 as the area of that setback space and keeps this area in mind while a subject considers himself or herself inside the setback space (i.e., between points x_1 and x_2). Thus, the total perceived area S^* can be obtained from the following equation:

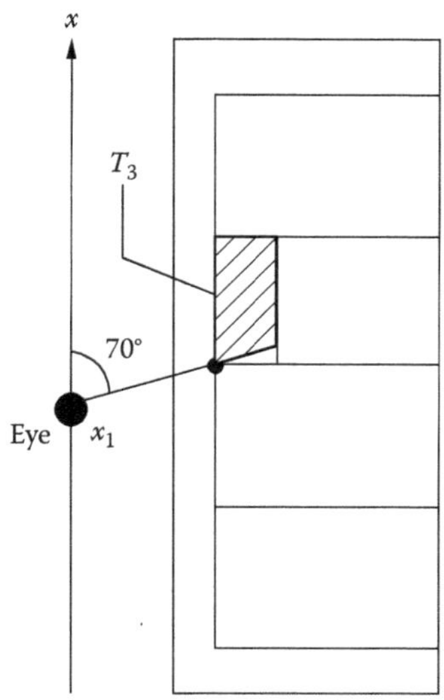

FIGURE 9.13 Largest visible area of a setback space.

TABLE 9.3
Total Perceived Area (S^*) and Estimated Values of Open-Space Ratios for Each Pattern

	Pattern					
	A	B	C	D	E	F
S^*	6.17	5.97	5.17	5.78	4.58	4.00
Estimate	0.31	0.26	0.08	0.04	–0.25	–0.45

$$S^* = \int_{x_1}^{x_2} S_3 dx \tag{9.4}$$

Because S_3 is a constant, Equation 9.4 is written as

$$S^* = \int_{x_1}^{x_2} S_3 dx = S_3 \left(x_2 - x_1\right) \tag{9.5}$$

By means of Equation 9.5, the total perceived area can be calculated for Patterns A to F, respectively. The estimated values of the open-space ratios for those patterns were obtained from the experiment (see Table 9.2). Table 9.3 shows the estimated values of the open-space ratios and the computed total perceived area for those patterns, and Figure 9.14 illustrates the relationship between these two variables. The correlation coefficient between the two is .96, which is significant at the .01 level.

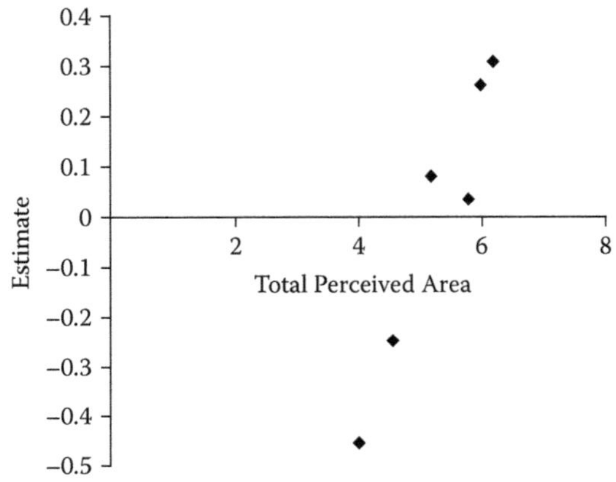

FIGURE 9.14 The relationship between the total perceived area and the estimated values of the open-space ratios. $r = .96, p < .01$.

Thus, Model 3 shows a very good fit to the observed data. From this result, it can be concluded that Model 3 is useful for explaining the perception of area.

9.4 DISCUSSION

In the literature, many experimental studies dealing with the perception of area using figures drawn on a piece of paper can be found (e.g., Anastasi, 1936; Holmberg and Holmberg, 1969; Martinez and Dawson, 1973). Sadalla and Oxley (1984) experimentally examined the perception of area at the scale of a building. They examined the perception of the area of a room and discussed perceptual characteristics in terms of the length/width ratio.

Although many studies examined the perception of area using figures drawn on a piece of paper or at the scale of a building, very few studies have been done at a city scale. Noticing this lack in research, this study investigated the perception of area with respect to open space in a city. One of the difficulties in conducting experiments of spatial perception is to control factors of concern. To overcome this difficulty, this study used a desktop virtual environment instead of conducting an experiment in an actual city.

As mentioned previously, in recent years, some studies have been conducted with a computer-simulated environment. The method used in this study is similar to those, but a distinction exists in that subjects in this experiment had a 3-D view through polarizing glasses and they walked through a simulated city. There is a great difference between static graphics and 3-D stereo dynamic graphics used here.

It was anticipated that the perception of the area of an open space would vary according to the shape and location of the space, even if the physical area is the same. Hence, the study focused on the shape and location of an open space and analyzed the effects of these factors on the perception of the area. From the experiment, two significant effects were found: the width/depth ratio effect and the visible area effect. The width/depth ratio effect is consistent with the results obtained by Sadalla and

Oxley (1984). Their results suggested that the area of a room was perceived as larger as the length/width ratio increased. In the present study, an open space with a wide frontage was perceived as larger than an open space with a deep setback. Regarding the width/depth ratio effect, similar results have also been found in studies dealing with figures on a piece of paper (e.g., Anastasi, 1936; Holmberg and Holmberg, 1969). These findings suggest that this effect might apply to various scales. Further research is needed to make an explicit statement about the generalizability. Also, the visible area effect is generally in line with the results obtained in the isovist analysis by Benedikt and Burnham (1985). In their study, the visible area of a space affected the estimation of spaciousness, although it dealt with only a static view from a specific vantage point. Obtained implications for the difference between a static view and a dynamic view in relation to perception models are discussed later in this chapter.

To explain the previously mentioned two effects (the width/depth ratio and the visible area effect), three perception models were formulated. Models 1 and 2 did not show a significant correlation with the observed data at the .05 level. These two results indicate that the total perceived area cannot be represented by merely integrating the visible area at every point. Thus, Model 3 was developed by introducing two notions: inside and outside of a setback space. A subject keeps the largest visible area of the setback space in mind while he or she considers himself or herself inside a setback space. It is interesting to notice that the notions of inside and outside can be applied to the perception of area. This model suggests that the results of the isovist analysis by Benedikt and Burnham (1985) need some modification when applied to the perception of area in a dynamic view. That is, the largest visible area, concurrent with the judgment of the inside and outside, affects perceived area, rather than merely a visible area at each vantage point.

This study has some limitations, however. First, it was conducted under the condition that subjects faced straight ahead and did not turn their heads around during the walk. In fact, while people walk in a city, they turn their heads and look around from time to time. This created a hidden portion of an open space in this experiment that could have been seen if the subjects had turned their heads around. Hence, this factor should be taken into account in a future study. Adding to that, movement of a head is generally said to have an effect on the perception of depth (studies of motion parallax) (e.g., Rogers and Graham, 1979). Second, this experiment only dealt with the shape and location of an open space. There are some other possible factors that may affect the perception of an area, such as the height of a building and the ratio of the street width to the building height. Third, the applicability of the results obtained here in a computer-simulated environment to an actual city should be discussed. The subjects in this experiment lacked information available from actual movement through an environment (e.g., optic flow, proprioception, sense of scale). Whether this causes a difference in an important way is one of the major topics for future research.

In spite of the above limitations, the results obtained from this study are suggestive for the planning of residential areas. For instance, when a park is designed along a street in a crowded city, the results would suggest how to design the shape and location of the park. A good way to produce a larger perceived area of the park with a fixed area would be to use the widest possible frontage and locate the park so that people can see a larger area of it. With further examinations of the issues discussed earlier, more detailed implications for planning should be obtained.

ACKNOWLEDGMENTS

The authors thank the researchers in the Hitachi Central Research Laboratory who participated in the experiment and Tomoko Hatakeyama for her assistance in organizing the experiment. Thanks are also due to Robert Bechtel and Daniel Montello for their valuable comments on earlier drafts of the chapter. This chapter was originally published in *Environment and Behavior* 30 (1998): 216–34. Reproduced by permission of Sage Publications, Inc., Thousand Oaks, CA.

REFERENCES

Anastasi, A. 1936. The estimation of area. *Journal of General Psychology* 14:201–25.

Benedikt, M., and Burnham, C. A. 1985. Perceiving architectural space: From optic arrays to isovists. In W. M. Warren & R. E. Shaw (Eds.), *Persistence and change: Proceedings of the First International Conference on Event Perception* (pp. 103–14). Hillsdale, NJ: Lawrence Erlbaum.

Golledge, R. G., Dougherty, V., and Bell, S. 1995. Acquiring spatial knowledge: Survey versus route-based knowledge in unfamiliar environments. *Annals of the Association of American Geographers* 85:134–58.

Higuchi, T. 1975. *The visual and the spatial structure of landscapes.* Tokyo: Giho-do.

Holmberg, L., and Holmberg, I. 1969. The perception of the area of rectangles as a function of the ratio between height and width. *Psychological Research Bulletin* 9:1–6.

Holmberg, L., and Wahlin, E. 1969. The influence of elongation on the judgment of the area of certain geometrical figures. *Psychological Research Bulletin* 9:1–11.

Martinez, N., and Dawson, W. E. 1973. Ranking of apparent area for different shapes of equal area. *Perceptual and Motor Skills* 37:763–70.

O'Neill, M. J. 1992. Effects of familiarity and plan complexity on wayfinding in simulated buildings. *Journal of Environmental Psychology* 12:319–27.

Rogers, B., and Graham, M. 1979. Motion parallax as an independent cue for depth perception. *Perception* 8:125–34.

Sadalla, E. K., and Oxley, D. 1984. The perception of room size: The rectangularity illusion. *Environment and Behavior* 16:394–405.

Smith, J. R. 1969. The effects of figural shape on the perception of area. *Perception and Psychophysics* 5:49–52.

Tlauka, M., and Wilson, R. N. 1994. The effect of landmarks on route-learning in a computer-simulated environment. *Journal of Environmental Psychology* 14:305–13.

Tlauka, M., and Wilson, R. N. 1996. Orientation-free representations from navigation through a computer-simulated environment. *Environment and Behavior* 28:647–64.

Section III

Spatial Analysis

Yukio Sadahiro

The definition of spatial analysis varies across academic fields and even among individual researchers. In its widest sense it is the analysis of space itself and that of phenomena observed in space. Spatial analysis in a narrower sense consists of (1) data collection, (2) data management and manipulation, (3) analysis, and (4) modeling/planning. The third step gives the narrowest definition of spatial analysis. Following the second definition, this section briefly looks at each step of spatial analysis.

Spatial analysis usually begins with data collection. This has often been the central concern in spatial analysis, because it has always been very costly in time and money. At present, however, new technologies such as global positioning systems (GPS), mobile geographic information systems (GIS), and the integrated circuit (IC) tag system permit us to collect spatial data with more ease and less cost than a decade ago. Even individual human movement, from in-room movement to trip activities by cars and trains, can be captured by laser scanner and cell phone tracking.

Continuous spatial phenomena such as temperature and wind speed distributions cannot be inherently measured at every location where they are defined. They can be observed discretely only at sample locations, though new devices, as mentioned above, can reduce the sampling interval. Consequently, their value at other locations has to be estimated from observations, which is called spatial interpolation. As Shiode and Shiode state in Chapter 10, numerous interpolation methods have been proposed in the literature. Most of them consider a two- or higher-dimensional plane infinite space, assuming that spatial phenomena can be defined and observed anywhere in such a space. Though it holds for natural phenomena, human activities are mainly performed only on a network space. If we take a train, we can move only on a railway network. Car driving is usually permitted on traffic roads. Spatial phenomena related to human activities are often limited to such a network space. To treat these cases, Shiode and Shiode develop a new method of spatial interpolation on a network space. They propose a theoretical framework of spatial interpolation on a network, which is followed by a computational procedure of interpolation using the inverse distance-weighted method. They applied the method as well as its counterpart in a planar space to interpolation on a network space to evaluate the accuracy of both methods.

Once spatial data are collected, they are managed in a GIS environment. Graphical user interface allows us to visualize the data easily and interactively in various cartographic forms. Visual analysis is effective especially at an early stage of spatial analysis.

Spatial analysis in its narrowest sense is often called exploratory spatial analysis. As seen in visual analysis, exploratory spatial analysis is performed to find interesting spatial patterns to build research hypotheses. It covers a wide range of spatial phenomena from point objects to continuous surfaces. Spatial tessellation, a set of polygons that fully cover a region without overlapping, also appears in spatial analysis. In the real world it represents census tracts, postal zones, market areas, land cover, soil patterns, and so forth.

Spatial tessellations are often closely related to each other. School districts and postal zones are based on census tracts and administrative units. Market areas of different categories of shops affect each other. To discuss such relations, Sasaya and Sadahiro in Chapter 11 propose a new mathematical method for analyzing the relations among spatial tessellations. The method considers three aspects of similarity between tessellations: granularity, structure, and hierarchy. They are evaluated by quantitative measures, which are used in the classification of tessellations and visualization of the relations among tessellations. To test the validity of the method, they applied it to the analysis of hypothetical spatial tessellations and candidate plans for a new administrative system called Doshusei in Japan. Though the method was originally developed for spatial tessellations, they suggest its extension to the analysis of other types of spatial distributions, such as polygon distributions.

Analysis of spatial phenomena greatly helps us in finding spatial patterns. It also reveals their underlying structure, which leads us to spatial models, quantitative representations of spatial phenomena. Spatial models mathematically describe spatial phenomena in either a deterministic or probabilistic form that permits us to predict their future status.

Another direction taken after spatial analysis is spatial planning, that is, location planning of spatial objects. In urban planning it represents the planning of facility location, touring routes, land use regulation, and so forth.

A powerful mathematical tool for spatial planning is spatial optimization. The last two chapters of this book use spatial optimization frameworks to discuss desirable spatial structures in urban areas.

Spatial optimization is one category of mathematical programming. It is generally unsolvable in the sense that the optimal solution cannot always be found in a practical computing time. Instead of the optimal solution, we usually seek a desirable solution by using heuristic methods.

In Chapter 12, Hori and Yoshikawa take the GA (genetic algorithm) approach, a heuristic method widely applicable to problems in mathematical programming.

GA works efficiently even for unsolvable problems, detecting desirable solutions in a practical time. As Hori and Yoshikawa mention, GA has been widely used in spatial optimization, from facility location to route planning. However, there are few studies that adopt GA in spatial location-allocation problems, simultaneous optimization of location facility, and demand allocation. To fill the gap of research, Hori and Yoshikawa propose a new method of GA to treat such a location-allocation problem. They first apply the method to a small problem defined in a hypothetical city to capture the properties of the method. To test the validity of the method, they proceed to an actual location-allocation problem in a suburban area of Tokyo.

While Hori and Yoshikawa discuss the optimal location of facilities, Okunuki considers the optimal system of administration in Chapter 13. A focus is on the efficiency of hierarchical administrative systems, from a mono-level system where all the public services are provided by the central government to a multi-level system where different services are provided by different levels of local governments. Assuming a one-dimensional hypothetical city, Okunuki calculates the minimum number and distribution of employees necessary for maintaining an administrative system. The resultant employee distribution is discussed in comparison with population agglomeration in the real world.

This section shows state-of-the-art methods and applications at different steps of spatial analysis. Though it does not cover the whole range of spatial analysis, it at least conveys a "taste" and "joy" of analyzing spatial phenomena.

10 Inverse Distance-Weighted Interpolation on a Street Network*

Shino Shiode and Narushige Shiode

10.1 INTRODUCTION

Spatial interpolation, or smoothing technique, is a method used for estimating an unknown spatial value using information of known values observed at a finite number of sample locations within the study area. It is applied in many instances to help improve the understanding of spatial or temporal phenomena by estimating unknown values at specific locations, or by generating a continuous surface that covers an entire study area. Spatial interpolation is considered to be one of the most fundamental spatial analytical operations, as it has found applications in a wide variety of research fields, including climatology, geostatistics, geomorphology, oceanography, and environmental study. A variety of spatial interpolation methods exist, each of which provides a good prediction under different estimation criteria. In recent years, these methods have become an integral part of geographic information systems (GIS) and other spatial analytical software.

The conventional spatial interpolation methods have one critical limitation in that they assume that both the sample locations and the target locations exist in the two-dimensional or three-dimensional Euclidean space, i.e., the distance between the samples and the targets is measured in a straight line. Although this assumption holds in most cases, there are phenomena that are observed or measured in other types of spaces, for instance, on a network, and which should therefore be analyzed in network space. Examples of such phenomena include water quality in rivers and waterways, the readouts along a power line, and medical indices from the circulatory and nervous systems of the human body. There are also other phenomena that can be measured in the Euclidean space but would benefit from increased accuracy if their attribute values were estimated along a network, e.g., the elevations of a terrain surface, the appraisal value of estate properties, and the

---

* This chapter has been adapted with updates and modifications from the following chapter: Shiode, S., "Inverse Distance Weighted Method for Point Interpolation on a Network," *Theory and Applications of GIS* 13 (2005): 33–41 (in Japanese).

gas emission level along the highways. Several studies suggest that application of network-based methods, as opposed to their conventional counterparts that assume the Euclidean space, can provide a better understanding of the phenomena observed on a network, as they account for the actual shortest-path distance among locations in the network space (Clevenger et al., 2003; Spooner et al., 2004; Yamada and Thill, 2004; Maheu-Giroux and de Blois, 2007; Okabe et al., 2008; Shiode and Shiode, 2009). It is also conceivable that a spatial interpolation method defined on a network can provide a more accurate estimation of network-based phenomena.

This chapter thus proposes a spatial interpolation method for estimating unknown values at locations on a network. It is an extension of an existing interpolation method and will be adapted to the network space. Section 10.2 provides a brief overview of the conventional, Euclidean space–based spatial interpolation methods. This is followed by an introduction to the network-based method and a description of how it can be used to give estimates along a network. Section 10.3 examines the validity and level of accuracy of the method through a comparative study of the network and the ordinary interpolation methods using two different types of street network data with elevation values measured along the streets. Section 10.4 brings the chapter to a conclusion with a discussion of findings and future directions.

10.2 METHODOLOGICAL FRAMEWORK OF SPATIAL INTERPOLATION ON A NETWORK

10.2.1 Spatial Interpolation in the Euclidean Space

Although a number of spatial interpolation methods have been developed for estimating values in the Euclidean space, it is impossible to identify one particular method that can constantly provide the best results under different estimation criteria. It is usually up to the users to select the method suitable for their particular needs. This section first reviews the existing range of interpolation methods with a focus on their methodological characteristics; it then discusses their suitability for adaptation to the analysis of phenomena on a network.

The existing range of interpolation methods can be divided into two groups, global and local methods, and the latter can be further classified into the continuous and the discrete. In general, the global methods are used for generating a continuous surface by taking the attribute values from all sample locations. Methods that fall in this category include the trend surface, and interpolation by the Fourier series. The global type is known to be unsuitable for estimating small-scale, local variations in the interpolated curve, as one small deflection of the observed value at a single sample location may affect the entire curve (Maze and Takeda, 2001).

The local interpolation, on the other hand, only uses the nearby samples to estimate the unknown spatial value. This makes the local methods a more suitable candidate for predicting in the network space, as most phenomena observed on a network tend to be affected by the local variations rather than the global trend or a periodic change on the global scale; i.e., unknown spatial values in network space are likely to inherit the attributes of their nearby samples and not those of far-distant locations.

Among the interpolation methods that are usually considered to be local are the various types of kriging (Matheron, 1963), proximal method (Peucker and Chrisman, 1975), natural neighbor (Sibson, 1981), inverse distance-weighted method (IDW) (Bailey and Gatrell, 1995), Clough–Tocker interpolation (Clough and Tocker, 1965; Silverman, 1981), and triangulated irregular network (TIN) and spline curves (Bartels et al., 1987). Strictly speaking, the boundary between the global and local methods is slightly ambiguous. This is because, in the case of the proximal method, natural neighbor, and TIN, the number of sample locations is predetermined by the computational procedure, whereas IDW and kriging can take any arbitrary number of samples. This means that, in theory, they could take on the entire set of samples across the study area, thus presenting a special case of global estimation. In reality, however, it would be impractical to use such a large number of samples, as it would only add more noise and lower the prediction accuracy; this is why IDW and kriging commonly use a small set of samples, and are hence regarded as local methods.

The local interpolation methods listed above can be further divided into two groups: the continuous and the discrete. For instance, Clough–Tocker interpolation and spline curves belong to the former, since they utilize a smooth curve that is fitted to the samples. The rest are categorized as the discrete type, where each of the unknown spatial values is predicted individually from its respective nearby sample locations. The discrete methods have a much wider variation, allowing us to look for an appropriate method from a wider pool. In addition, their computational load is relatively light, as they only require a small number of samples. And it is for these tractable features for dealing with the real-world phenomena that the local, discrete methods have been most widely used.

With the exception of TIN, all the discrete methods listed above predict the unknown spatial values as a function of the weighted mean of a fixed number of samples. In that sense, they can be considered as a variation of the inverse distance-weighted method (IDW). In other words, IDW offers a foundation from which various other discrete interpolated methods can be developed. Considering the versatility of the methods that utilize IDW, this chapter proposes an IDW interpolation method for the network space, which can be used in future studies as a foundation for the network variation of other interpolation methods.

10.2.2 Inverse Distance-Weighted Method for a Point Interpolation on a Network

Figure 10.1 shows a small portion of the street network in Tochigi, Japan. The 11 points p_1, ... p_{11} in the figure indicate the sample locations where the observed values were measured. Let us consider predicting the elevation level at a location where the altitude is unknown, using the observed values at sample locations on the street network of the study area.

Let p_A be the target location with an unknown value to be predicted. When a discrete and local interpolation method is applied, a fixed number (typically between

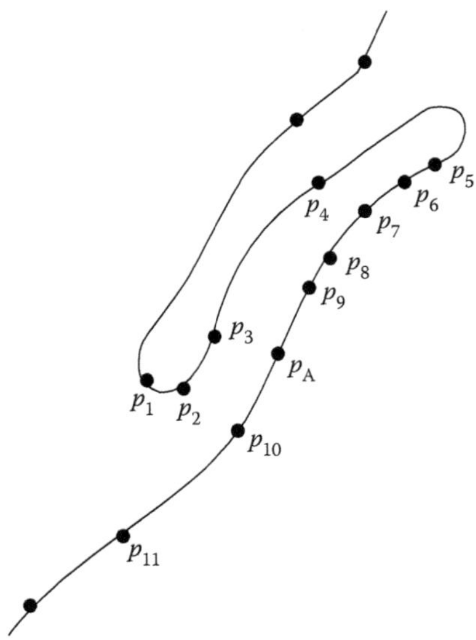

FIGURE 10.1 An example of a street network (ridge line) with spot heights as elevation points.

two and eight) of sample points around the target needs to be determined. These points are hereafter called nearby sample locations.

Suppose that, based on their straight-line distance from p_A, the six nearest points are to be identified as the nearby sample locations for predicting the elevation at p_A. In the case of Figure 10.1, the six sample locations are p_1, p_2, p_3, p_8, p_9, and p_{10}. Among the sample locations, p_3, p_9, and p_{10} are closer to p_A than the rest, if measured in straight-line distance on a Euclidean plane. If, however, we consider the shortest-path distance along the network (Aho et al., 1983; Dijkstra, 1959), p_3 is in fact located much farther away from p_A.

In general, elevation levels are continuous along a street network. In contrast, if two locations are found close to each other but without any straight street segment in between, e.g., p_A and p_3, the elevation levels are not necessarily continuous. In other words, there may be a significant difference in their altitude, especially in a mountainous area. If the interpolation is conducted with data that contain such differences in elevation, it could undermine the level of prediction accuracy. In such cases, it would be preferable to identify the nearby sample locations with respect to their shortest-path distance from the target location.

Based on the shortest-path distance along the street network, the point that has the closest value to p_A in Figure 10.1 is unlikely to be p_3 but could be either p_9 or p_{10}, the closest sample locations to p_A along the network in opposite directions. The new set of the six nearest sample locations to p_A based on the shortest-path distance are p_6, p_7, p_8, p_9, p_{10}, and p_{11}. As this example illustrates, the combination of nearby sample locations is different between the planar-based and network-based norms, which lead to different estimates. This chapter aims to confirm that a spatial interpolation method defined in the network space produces more accurate prediction values than the conventional interpolation would in the Euclidean space.

Let us first define a method that predicts an unknown spatial value on a network using the observed values at nearby sample locations along the network and weight it with respect to the shortest-path distance from these sample locations. We will hereafter call such a method the network inverse distance-weighted method (NT-IDW), and distinguish it from the conventional IDW, which we will refer to as the planar inverse distance-weighted method (PL-IDW). The NT-IDW is carried out in three steps:

1. Find a fixed number of points closest to the target location by conducting a shortest-path distance search, and identify them as the nearby sample locations
2. Calculate the weight as an inverse power function of the shortest-path distance between each of the nearby sample locations and the target location
3. Predict the unknown spatial value as the weighted mean of the observed values at the nearby sample locations

Let p_0 be the target location on the network, and z_0 be the unknown value of p_0 to be interpolated. Let $p_1, p_2, \ldots, p_s$ be s number of nearby sample locations with respective observed values of $z_1, z_2, \ldots, z_s$. Then z_0 is predicted as the weighted mean of the observed values at the nearby sample locations:

$$\hat{z}_0 = \sum_{i=1}^{s} w_i z_i, \quad \sum_{i=1}^{s} w_i = 1 \tag{10.1}$$

where $W = w_i$ $(i = 1, 2, \ldots, s)$ is the relative weight assigned to the nearby sample locations. Let $d_{\mathrm{NT}}(p_0, p_i)$ $(i = 1, 2, \ldots, s)$ denote the shortest-path distance from a nearby sample location p_i $(i = 1, 2, \ldots, s)$ to the target location p_0 along the network. Weight w_i of p_i can be identified as a function of the shortest-path distance. Equation 10.1 can be rewritten as

$$\sum_{i=1}^{s} w_i z_i = \frac{\sum_{i=1}^{s} f(d_{\mathrm{NT}}(p_0, p_i)) z_i}{\sum_{i=1}^{s} f(d_{\mathrm{NT}}(p_0, p_i))} \tag{10.2}$$

In general, the influence of the observed value at the nearby sample locations on the unknown value at the target location is assumed to weaken as the distance between them increases. Taking the effect of distance decay into account, the inverse power function of d_{NT} is adopted to calculate the weight W, and $\hat{z}_0$ at p_0 is predicted as follows:

$$\hat{z}_0 = \frac{\sum_{i=1}^{s} d_{NT}(p_0, p_i)^{-\lambda} z_i}{\sum_{i=1}^{s} d_{NT}(p_0, p_i)^{-\lambda}} \tag{10.3}$$

where λ is the power coefficient.

There are two issues that need to be noted here. First, since the predicted value of $\hat{z}_0$ may change with parameters s and λ, identifying appropriate parameter values is crucial for successful prediction. Unfortunately, no method is known to determine the optimum parameter values (for instance, 2 is commonly adopted for λ to offset the distance decay, but Isaaks and Srivastava (1989) showed that it does not always result in a high degree of estimation accuracy). In this chapter, we will repeat the interpolation process using different combinations of λ and s to achieve more accurate prediction.

The other issue concerns the way the nearby sample locations are determined; they can be assigned by taking a predetermined number of nearby sample locations, or they can be based on search distance of a predetermined length. In the case of the latter, any sample locations found within the search distance would become the nearby sample locations, but the number of samples may vary from one target location to another. As it would be preferable to use the same number of nearby sample locations for all target locations and also between NT-IDW and PL-IDW for comparison, this chapter will adopt the former.

10.2.3 Inverse Distance-Weighted Interpolation in Planar and Network Spaces

Based on the method stated in the previous section, a computer program was developed to execute NT-IDW and PL-IDW simultaneously. The program has also been adopted as part of a comprehensive GIS tool, SANET Version 3, which was developed to support spatial analysis on networks (please refer to Okabe et al., 2006a, 2006b for more details).

In theory, any unknown spatial value on a network can be estimated by entering arbitrary values of λ and s to the program, and it would allow us to generate a sufficient number of target locations to construct a smooth 3D surface. While such a 3D surface may have a practical utility, it adds little to confirmation of the validity of the network-based interpolation method. For this reason, the interpolation we conduct in this chapter will consist of a small number of predicted locations that are primarily intended for validity testing.

One way to confirm the validity of the proposed method is to use a cross-validation technique (Griffith and Layne, 1999). In the cross-validation, each sample location can be treated as a target location with an unknown value; i.e., we assume that the observed attribute values are unknown to us and predict their value from their respective nearby sample locations. The predicted values obtained through interpolation are compared with the observed values to examine the accuracy of the interpolation method.

Suppose that there are n number of sample locations $p_1, p_2, \ldots, p_n$ in the study area, and that the observed values at each sample location are denoted by $z_1, z_2, \ldots, z_n$. Let the predicted attribute values for NT-IDW and PL-IDW be $\hat{z}_i^{NT}, \hat{z}_i^{PL}$ for target locations p_i $(i = 1, 2, \ldots, n)$. An index for comparing the prediction errors of $\hat{z}_i^{NT}, \hat{z}_i^{PL}$ from the observed value z_i can be defined as follows:

$$r_i^{NT} = \hat{z}_i^{NT} - z_i, \; r_i^{PL} = \hat{z}_i^{PL} - z_i \;\; (i = 1, 2, \ldots, n) \quad (10.4)$$

Mean-squared errors (MSEs) (Isaaks and Srivastava, 1989) for the network-based and planar-based interpolations can be written as follows:

$$M_{NT} = \frac{1}{n}\sum_{i=1}^{n}(r_i^{NT})^2, \; M_{PL} = \frac{1}{n}\sum_{i=1}^{n}(r_i^{PL})^2 \quad (10.5)$$

In addition to MSE, which is an aggregated index, we introduce another index, D, to show the difference in the degree of disparity of the prediction errors between NT-IDW and PL-IDW for each target location. The degree of disparity in prediction error at target location p_i can be written as

$$D_i = \left|r_i^{NT}\right| - \left|r_i^{PL}\right| \;\; (i = 1, 2, \ldots, n) \quad (10.6)$$

In other words, the D index helps to identify individually which sample location, and to what degree, confirms the performance of NT-IDW over PL-IDW in providing a better prediction during cross-validation. For instance, a negative value of the D index confirms that NT-IDW provides a more accurate prediction than PL-IDW does for that particular location.

10.3 COMPARISON OF THE NETWORK AND THE PLANAR INTERPOLATIONS

10.3.1 DATASETS

In order to perform the cross-validation, we adopt two sets of elevation data measured on a network. In order to appreciate the impact of the street network configuration on the result, we prepare the following two types of network datasets.

Data 1 is a street network with 104 sample locations that follows a relatively simple but winding ridge line in a study area of 400 m by 450 m, whose elevation level ranges from 28 m to 84 m (Figure 10.2). In contrast, Data 2 consists of a road network with 148 sample locations that follows a denser, grid-like configuration that extends to a study area of 500 m by 600 m in a downtown district, whose elevation level is confined between 18 m and 34 m (Figure 10.3). In both cases, the street segments in the original networks extend beyond the study area.

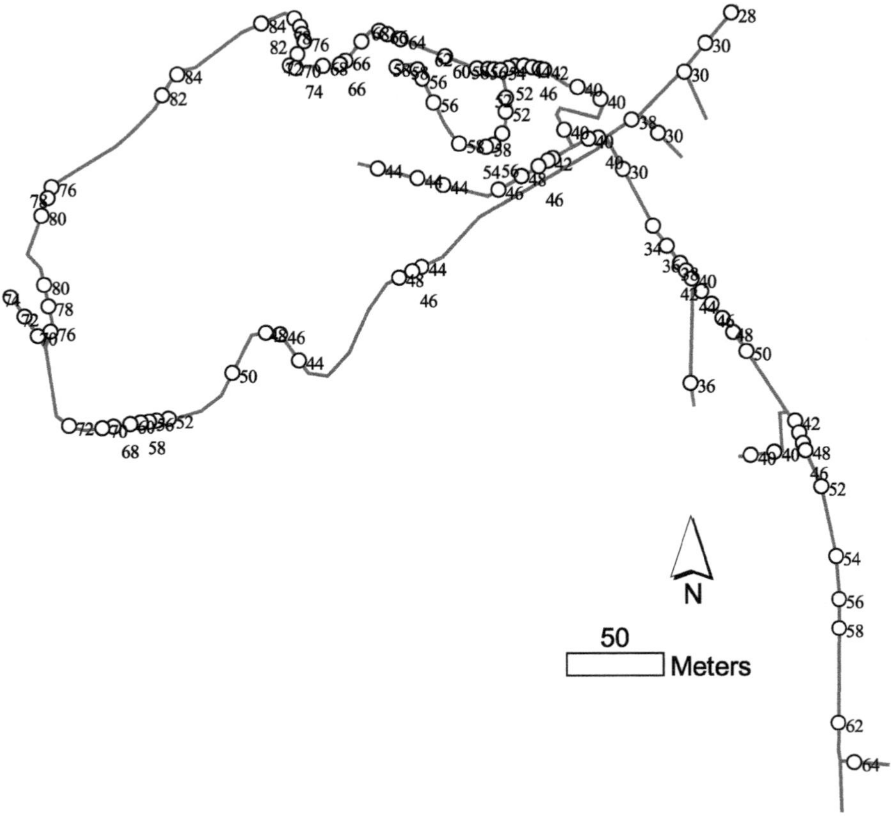

FIGURE 10.2 Data 1: A network containing 104 sample locations. The elevation at each location is shown in meters.

10.3.2 Interpolated Results from Data 1

Figure 10.4 provides a 3D representation of the result of the interpolation constructed with $s = 5$ and $\lambda = 1$. The amount of extrusion underneath the street network indicates the predicted elevation level at each polyline node.

Table 10.1 shows the result of the cross-validation calculated at all sample locations in Data 1 with the parameters taking a range of values at $s = 2, 3, \ldots, 8$, $\lambda = 0.5, 1, \ldots, 3$. The values adopted by s cover the number of nearby sample locations determined by the computation geometrical processes of frequently used methods such as TIN and natural neighbor.

The upper section of each cell shows MSE of NT-IDW, and the lower section shows MSE of PL-IDW. Of the two MSEs in each cell, the smaller value is underscored to indicate the one with an estimate closer to the observed value (Table 10.1). The MSEs confirm that, overall, NT-IDW maintains a higher degree of prediction accuracy than PL-IDW does for all combinations of s and λ. In addition, the D index measures the degree of disparity between the prediction errors from the two methods at each individual location. In Figure 10.5, the sample locations at which

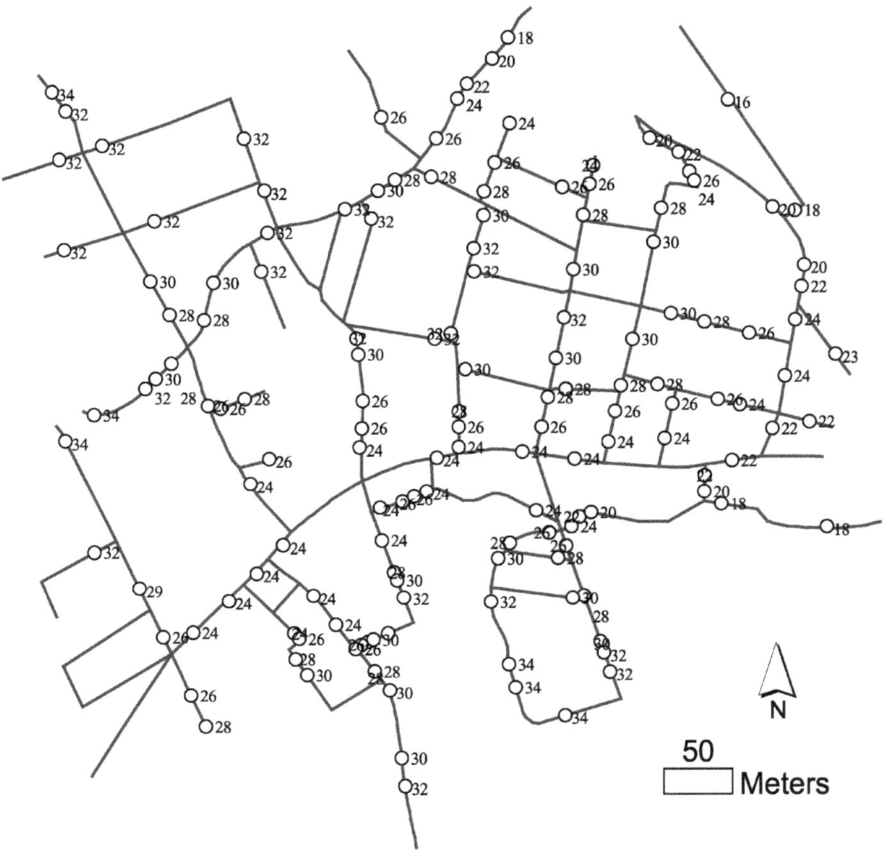

FIGURE 10.3 Data 2: A network containing 148 sample points. The elevation at each location is shown in meters.

NT-IDW outperformed PL-IDW are shown in black circles. Note that they are on a long stretch or winding network segments, which make the shortest-path distance to other locations much longer than the straight-line distance. This means that the difference in the combination of nearby sample locations for NT-IDW and PL-IDW, especially at locations marked by the black circles, leads to a considerable amount of difference in the interpolated elevation values.

10.3.3 Interpolated Results from Data 2

In order to explore the prediction accuracy of NT-IDW for a different network configuration, the same procedure is carried out with Data 2. First, we predict the elevation level of the polyline nodes to generate a 3D extrusion of the street network, as shown in Figure 10.6.

Second, using parameters $s = 2, 3, \ldots, 8$, $\lambda = 0.5, 1, \ldots, 3$, cross-validation is conducted across the entire set of sample locations. The results are summarized in Table 10.2, where the two values in each cell indicate MSEs of NT-IDW and PL-IDW.

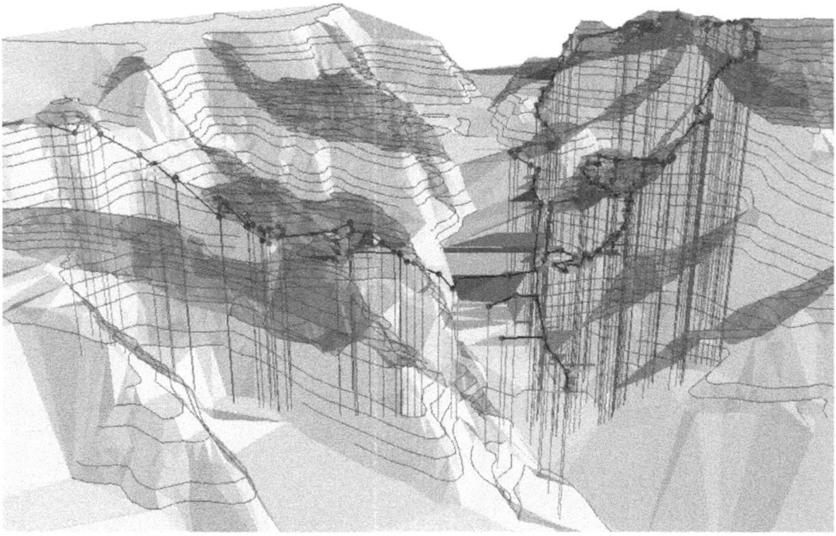

FIGURE 10.4 3D representation of the street network from Data 1 estimated with NT-IDW at $s = 5$ and $\lambda = 1$ (with 3× vertical exaggeration).

TABLE 10.1
MSEs of the Planar and the Network Interpolations of Data 1

λ \ s	2	3	4	5	6	7	8
0.5	**5.924** 7.874	**6.236** 9.381	**7.323** 10.767	**8.977** 12.919	**11.012** 14.701	**12.104** 16.512	**13.916** 17.252
1	**5.479** 7.449	**5.484** 8.481	**6.265** 9.173	**7.178** 10.512	**8.429** 11.562	**9.019** 12.560	**10.152** 13.089
1.5	**5.162** 7.174	**4.991** 7.843	**5.529** 8.070	**5.968** 8.840	**6.698** 9.393	**6.956** 9.866	**7.595** 10.161
2	**4.951** 7.019	**4.689** 7.413	**5.062** 7.372	**5.232** 7.790	**5.651** 8.049	**5.733** 8.252	**6.082** 8.403
2.5	**4.818** 6.950	**4.514** 7.136	**4.781** 6.951	**4.813** 7.161	**5.059** 7.259	**5.062** 7.338	**5.256** 7.416
3	**4.742** 6.938	**4.422** 6.967	**4.623** 6.710	**4.589** 6.799	**4.739** 6.808	**4.713** 6.832	**4.826** 6.872

* The upper section of each cell shows MSE of the network interpolation, and the lower section shows MSE of the planar interpolation.

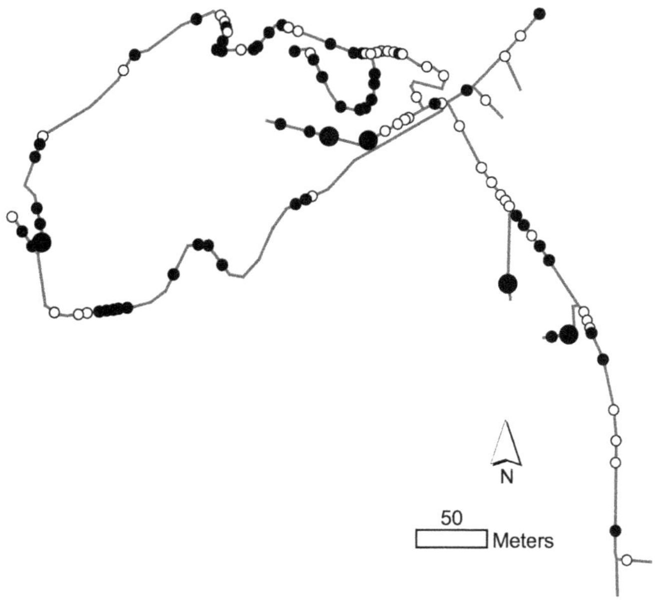

FIGURE 10.5 Illustration of the relative prediction accuracy of NT-IDW at each individual location in Data 1. Target locations at which NT-IDW outperformed PL-IDW are shown in black circles. Small circles show those with an average D index value between –2 and 0, and large circles show locations with an average D index smaller than –2 (i.e., NT-IDW achieved much higher prediction accuracy).

TABLE 10.2
MSEs of the Planar and the Network Interpolations of Data 2

λ \ s	2	3	4	5	6	7	8
0.5	**2.345**	2.747	3.276	3.532	3.870	4.091	4.414
	2.504	**2.530**	**2.920**	**3.161**	**3.604**	**3.655**	**3.806**
1	**2.256**	2.514	2.886	3.055	3.282	3.430	3.645
	2.394	**2.367**	**2.632**	**2.783**	**3.103**	**3.121**	**3.258**
1.5	**2.233**	2.405	2.654	2.757	2.898	2.989	3.121
	2.349	**2.296**	**2.466**	**2.541**	**2.759**	**2.741**	**2.848**
2	**2.250**	2.371	2.539	2.600	2.686	2.740	2.819
	2.347	**2.282**	**2.394**	**2.420**	**2.567**	**2.528**	**2.606**
2.5	**2.288**	2.377	2.492	2.529	2.581	2.615	2.665
	2.369	**2.300**	**2.378**	**2.378**	**2.482**	**2.437**	**2.494**
3	**2.333**	2.401	2.483	2.506	2.537	2.560	2.592
	2.403	**2.334**	**2.394**	**2.380**	**2.456**	**2.415**	**2.456**

* The upper section of each cell shows MSE of the network interpolation, and the lower section shows MSE of the planar interpolation.

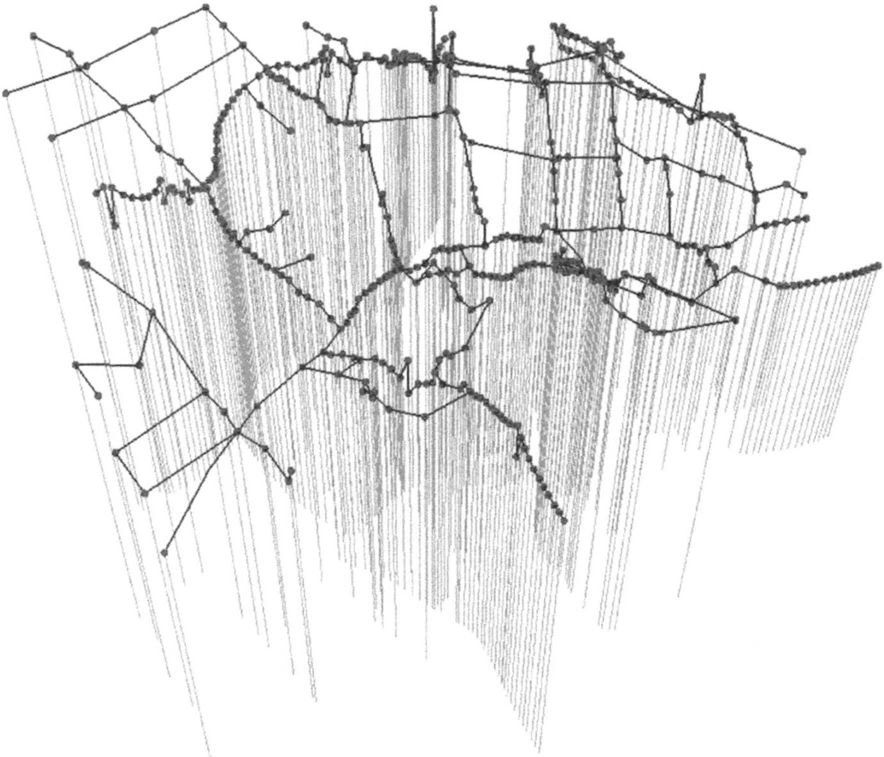

FIGURE 10.6 3D representation of the street network from Data 2 estimated with NT-IDW at $s = 5$ and $\lambda = 1$ (with 10× vertical exaggeration).

Note that the difference between each pair of MSEs is much smaller than that of Data 1. Once again, within each cell, the one with the smaller MSE value, or the one with a predicted value closer to the observed value, is underscored. The table clearly shows that NT-IDW was only successful in providing higher prediction values at $s = 2$, and that PL-IDW outperformed NT-IDW in all other cases. Figure 10.7 shows a comparison of the accuracy of the predicted values for each individual location between NT-IDW and PL-IDW through the mean value of their D index ($s = 2, 3, \ldots, 8$, $\lambda = 0.5, 1, \ldots, 3$); black points indicate locations where NT-IDW gave prediction values of low accuracy. As these points are mostly found on the periphery of the network, it is assumed that the edge effect of the network (Ripley, 1981) may be involved in this case.

In other words, the edge effect is likely to be a main source of prediction error. The true nearby sample locations for each target location are not completely identified in either NT-IDW or PL-IDW, as some of them may be outside the study area. In addition, in the case of NT-IDW, some of the sample locations are excluded from the set of nearby sample locations even though they are within the study area. This is because of the discontinuity of the street network in the periphery of the study area, which prevents us from measuring the shortest-path distance from the nearby sample locations to the target location. In contrast, in the case of PL-IDW, all nearby sample

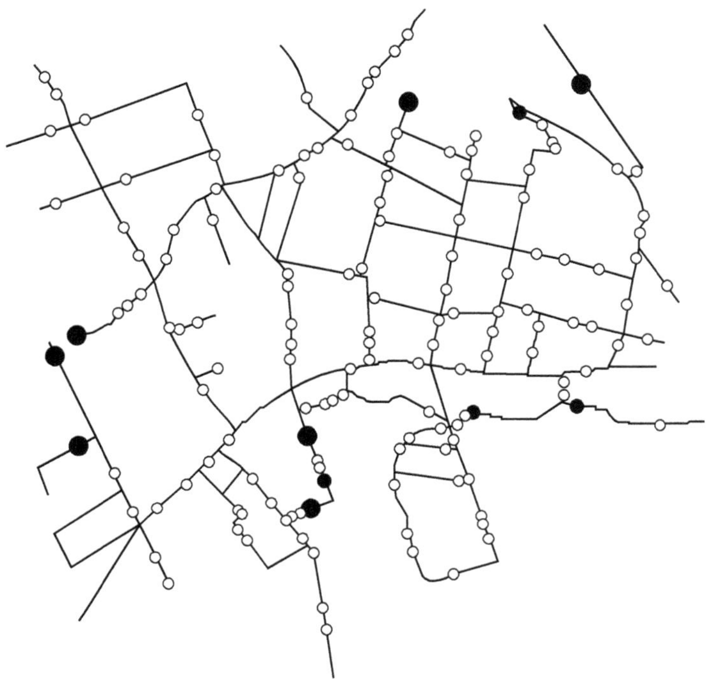

FIGURE 10.7 Illustration of the relative prediction accuracy of NT-IDW at each individual location in Data 2. Target locations at which NT-IDW outperformed PL-IDW are shown in white circles, and those at which NT-IDW underperformed are shown in black. Small circles show those with an average *D* index value between 3 and 5, and large circles show locations with an average *D* index greater than 5 (i.e., NT-IDW gave a particularly poor estimate).

locations can be properly identified as long as they are inside the study area, because the distance from the sample locations to the target location is measured in straight-line distance, and is not affected by the discontinuity of the street network on the periphery of the study area. The lower degree of prediction accuracy by NT-IDW can be thus attributed to the edge effect, which excludes the sample locations outside the study area and also disrupts the street network in the periphery of the study area.

10.3.4 Interpolated Results from Data 2 after Modification

In order to eliminate the edge effect and to conduct the cross-validation of NT-IDW and PL-IDW under equal terms, the following modification is made on Data 2. As Data 2 is part of larger data covering a wider study area, it can be modified by expanding the study area outwards. Note that a series of edge correction methods exist to treat the cases where the information on the areas outside the study area is not available: they include toroidal edge correction (Ripley, 1988), translation correction (Ohser and Stoyan, 1981), isotropic correction (Ripley, 1976) and guard area method (Cressie, 1993). As they are primarily intended for application in the Euclidean space, they cannot be utilized for the edge correction in a network space without

FIGURE 10.8 Network data from Data 2 after modification; 104 sample locations are added to the original sample locations of 148 points for edge correction (the newly extended portion of the network outside the original study area is shown in thin line segments, and the additional sample locations in triangles).

theoretical adjustments. Among these existing methods of edge correction, the guard area method would be the most appropriate candidate for adoption in the network space, as it creates a network-based buffer zone at the periphery of the study area.

Fortunately, for Data 2, we have the actual street network data of the areas immediately outside the original study area, and the process of data modification can be thus completed in two steps:

1. Add network segments outside the study area and continue to expand them until all nearby sample locations for every target location are accounted for.
2. Add the sample locations on the newly added network segments to the set of sample location candidates. Note that these locations will only be used as sample locations and will not be added to the set of target locations.

Figure 10.8 shows the network data from Data 2 after modification, where 104 sample locations are added to the original 148 sample locations (the newly extended portion of the network outside the original study area is shown in thin line segments, and the additional sample locations in triangles).

FIGURE 10.9 Increase in the level of prediction accuracy of NT-IDW at each individual location of Data 2 after modification. Sample locations at which the prediction accuracy showed improvement are marked in black circles (large circles: high accuracy with average D index greater than 3; small circles: average D index between 1 and 3).

Table 10.3 summarizes the results from the cross-validation of Data 2 after modification. There are two notable differences from Table 10.2. First, the overall prediction accuracy is improved for both NT-IDW and PL-IDW. Second, NT-IDW predicts more accurately than PL-IDW does for all combinations of parameters except for $s = 7, \lambda = 3$.

Once again, the prediction accuracy for each individual sample location is measured by the mean value of the D index for each target location under every combination of $s = 2, 3, \ldots, 8, \lambda = 0.5, 1, \ldots, 3$. Results are shown in Figure 10.9, where sample locations with improved prediction accuracy are illustrated in solid black circles. Judging from their locations, we notice that the improvement in prediction accuracy near the periphery of the original network has played a critical role in improving MSE of NT-IDW. This is because the combinations of the nearby sample locations are largely changed for NT-IDW, but not so much for PL-IDW. This result also shows that even though the predicted values are affected by the edge effect in both the planar and network cases, the impact is much greater in the network case.

TABLE 10.3
MSEs of the Planar and the Network Interpolations of Data 2 with Modification

λ \ s	2	3	4	5	6	7	8
0.5	**1.926** 2.189	**2.261** 2.325	**2.366** 2.514	**2.324** 2.629	**2.453** 2.929	**2.699** 2.934	**2.744** 2.986
1	**1.881** 2.127	**2.097** 2.194	**2.143** 2.305	**2.129** 2.359	**2.220** 2.572	**2.369** 2.546	**2.400** 2.618
1.5	**1.898** 2.125	**2.049** 2.148	**2.048** 2.200	**2.040** 2.199	**2.099** 2.340	**2.182** 2.285	**2.192** 2.355
2	**1.951** 2.158	**2.065** 2.152	**2.039** 2.173	**2.031** 2.134	**2.069** 2.226	**2.113** 2.156	**2.111** 2.216
2.5	**2.018** 2.208	**2.110** 2.182	**2.074** 2.191	**2.066** 2.129	**2.040** 2.199	**2.114** 2.119	**2.108** 2.170
3	**2.088** 2.264	**2.165** 2.224	**2.128** 2.230	**2.120** 2.156	**2.134** 2.199	2.149 **2.134**	**2.142** 2.175

* The upper section of each cell shows MSE of the network interpolation, and the lower section shows MSE of the planar interpolation.

10.4 CONCLUSIONS

This chapter proposed a network-based interpolation method to predict unknown spatial values along a network. It extended IDW interpolation, a discrete and local interpolation method designed for application in the Euclidean space, and utilized it for interpolation in the network space. The proposed method, NT-IDW, was put through a comparative study in the form of cross-validation by means of two different datasets. Empirical analysis of Data 1 suggests that, if the topological structure of the network is relatively simple, comprising a small number of network segments, and there is a large variation in the range of the observed values, the difference in the combinations of the nearby sample locations between the planar and the network cases would lead to a large disparity in the predicted values produced by the two methods, typically resulting in a much lower prediction accuracy from PL-IDW.

On the other hand, analysis of Data 2 suggests that, when the configuration of the network takes a more regular, grid-like shape, and if the density of the samples becomes higher, the combinations of the nearby sample locations for each target location for the planar and the network cases become similar. It also demonstrates that the difference in the predicted values between the planar and the network cases becomes smaller as the range of attribute values at sample locations gets smaller. Given these observations, NT-IDW is particularly effective when applied to a network configuration like Data 1, where the straight-line distance and the shortest-path distance can be significantly different, whereas a dense and grid-like network configuration such as Data 2 can be analyzed with PL-IDW, as the difference between the straight-line distance and the shortest-path distance would be marginal. The study also showed that, even in the case of the grid-like network of Data 2, a higher level of prediction accuracy can be achieved after the data were modified to provide nearby sample locations for NT-IDW and PL-IDW.

As the results from the analysis on Data 2 indicate, NT-IDW is likely to be more susceptible to the edge effect, which is caused by the disruption of the network along the periphery of the study area. While the edge effect also affects Data 1, the amount of network disruption is far less than that in the case of Data 2, and this is clearly reflected by the overall high accuracy of prediction. In fact, NT-IDW was also applied to Data 1 with modification on their edges, which only made a slight improvement in the prediction accuracy.

Since the edge effect inevitably affects every dataset except in very special cases where the network is self-complete within the study area, it is always preferable to modify the network data with an appropriate edge correction procedure. Furthermore, prediction accuracy varies according to the location of the target and the sample locations as well as the size and the configuration of the network in the study area. As the empirical studies on the two datasets do not provide sufficient ground to offer more insights into the validity of NT-IDW, further testing with other datasets is desirable. Once its validity as an interpolation method has been confirmed, it may still require further modification to improve the prediction accuracy.

The method proposed in this chapter marks the first attempt at developing a network-based interpolation method, and it can be extended in several different directions. These include the treatment of vertical displacement and exploration of different combinations of parameters.

In terms of applications, elevation data used in this study are merely an illustrative example of phenomena observed on a network, and the network interpolation method can be applied to a range of other phenomena. For instance, the accuracy of real estate appraisal may improve by combining a network interpolation method with hedonic models to account for the exogenous growth. Adding these factors as parameters in the prediction process would improve the prediction accuracy of the interpolation at an unknown location.

In addition, the proposed method only utilizes IDW, which is one of the many existing planar interpolation methods. Using NT-IDW as a foundation, other variations of IDW-type interpolation methods may be adopted for application in the network space. Finally, certain types of applications (e.g., a detailed contouring of a bike path in a mountain range) may require the development of a network-based continuous interpolation method for generating smooth network paths.

ACKNOWLEDGMENTS

The authors greatly appreciate the input from Professor Atsuyuki Okabe in the early stages of the chapter. This study was supported in part by Grants-in-Aid for Scientific Research, Ministry of Education, Culture, Sports, Science and Technology, Japan.

REFERENCES

Aho, A., J. Hopcroft, and J. Ullman. 1983. *Data structure and algorithms*. Reading, MA: Addison-Wesley.

Bailey, T., and A. Gatrell. 1995. *Interactive spatial analysis*. London: Longman.

Bartels, R. H., J. C. Beatty, and B. A. Barsky. 1987. *An introduction to splines for use in computer graphics and geometric modelling*. Los Altos, CA: Morgan Kaufman.

Clevenger, A. P., B. Chruszcz, and K. E. Gunson. 2003. Spatial patterns and factors influencing small vertebrate fauna road-kill aggregations. *Biological Conservation* 109:15–26.

Clough, R., and J. L. Tocker. 1965. Finite element stiffness matrices for analysis of plates in bending. Paper presented at Proceedings of the 1st Conference on Matrix Methods in Structural Mechanics, Wright-Patterson AFB.

Cressie, N. 1993. *Statistics for spatial data*. New York: Wiley.

Dijkstra, E. 1959. A note on two problems in connection with graphs. *Numerische Mathematik* 1:269–71.

Griffith, D. A., and L. J. Layne. 1999. *A casebook for spatial statistical data analysis*. New York: Oxford University Press.

Isaaks, E. H., and R. M. Srivastava. 1989. *Applied geostatistics*. New York: Oxford University Press.

Maheu-Giroux, M., and S. de Blois. 2007. Landscape ecology of *Phragmites australis* invasion in networks of linear wetlands. *Landscape Ecology* 22:285–301.

Matheron, G. 1963. Principles of geostatistics. *Economic Geology* 58:1246–66.

Maze, S., and J. Takeda. 2001. *Spatial data modeling*. Tokyo: Kyoritsu-Shuppan.

Ohser, J., and D. Stoyan. 1981. On the second-order and orientation analysis of planar stationary point processes. *Biometrical Journal* 23:523–33.

Okabe, A., K. Okunuki, and S. Shiode. 2006a. The SANET toolbox: New methods for network spatial analysis. *Transactions in GIS* 10:535–50.

Okabe, A., K. Okunuki, and S. Shiode. 2006b. SANET: A toolbox for spatial analysis on a network. *Geographical Analysis* 38:57–66.

Okabe, A., T. Satoh, T. Furuta, A. Suzuki, and K. Okano. 2008. Generalized network Voronoi diagrams: Concepts, computational methods, and applications. *International Journal of Geographical Information Science* 22:965–94.

Peucker, T. K., and N. Chrisman. 1975. Cartographic data structures. *The American Cartographers* 2:55–69.

Ripley, B. D. 1976. The second-order analysis of stationary point processes. *Journal of Applied Probability* 13:255–66.

Ripley, B. D. 1981. *Spatial statistics*. New York: Wiley.

Ripley, B. D. 1988. *Statistical inference for spatial processes*. Cambridge, MA: Cambridge University Press.

Shiode, S., and N. Shiode. 2009. Detection of multi-scale clusters in network space. *International Journal of Geographical Information Science* 23:75–92.

Sibson, R. 1981. A brief description of natural neighbor interpolation. In *Interpreting multivariate Data*, ed. V. Barnett, 21–36. New York: Wiley.

Silverman, B. W. 1981. Density estimation for univariate and bivariate data. In V. Barnett, ed., *Interpreting multivariate data*, 37–53. New York: Wiley.

Spooner, P., I. D. Lunt, A. Okabe, and S. Shiode. 2004. Spatial analysis of roadside *Acacia* populations on a road network using the network K-function. *Landscape Ecology* 19:491–99.

Yamada, I., and J.-C. Thill. 2004. Comparison of planar and network K-functions in traffic accident analysis. *Journal of Transport Geography* 12:149–58.

11 Analysis of the Similarity between Spatial Tessellations *Method and Application*

Toshinori Sasaya and Yukio Sadahiro

11.1 INTRODUCTION

Spatial tessellation is one of the most important spatial structures in spatial analysis. Census tracts, postal zones, and electoral and school districts are defined for administrative purposes. Others are based on natural phenomena and include land cover, vegetation, and soil patterns. Market areas of retail stores, drug stores, and gas stations, for example, are often approximated by a set of polygons such as Voronoi diagrams.

Spatial tessellations in the same region are often closely related to each other (Okabe et al., 2000; Sadahiro, 2002). School districts and postal zones are sometimes based on administrative units. Transportation analysis zones (TAZs) are determined by census tracts and administrative units. A close relationship exists between administrative units and land use patterns. Market areas of different categories of shops affect each other because of consumers' propensity for one-stop multipurpose shopping. It is therefore necessary to analyze the relationship between tessellations to understand not only the relationship itself, but also the individual tessellations and their underlying structure.

Comparison and analysis of spatial tessellations depend on the comparability of attributes represented by tessellations. If we have time-series data for land use patterns in the same region, and the data share the same land use categories, the tessellations can be compared in terms of either spatial or attribute properties. The difference in the land use category is evaluated as well as the geometrical properties of tessellations. Two tessellations are regarded as different even if they are geometrically identical but different in land use pattern. On the other hand, tessellations of different attribute types are comparable only in terms of spatial properties. The difference among census tracts, postal zones, and electoral districts can be evaluated only in terms of their spatial properties. School districts for elementary, secondary, and higher education are not comparable in their attributes even if they are in the same region. Tessellations are regarded as identical only if they are geometrically identical.

In the above cases, analysis starts with a visual comparison of tessellations. Visual analysis is a useful means of obtaining a research hypothesis about spatial phenomena (Nielson et al., 1997; Kraak, 2003; Slocum et al., 2004; Ware, 2004; Wright, 2006). For time-series analysis of tessellations, animated visualization is effective for capturing significant or interesting changes in tessellations.

Visual analysis is often followed by quantitative analysis. Statistical measures are available, including χ^2, the Kappa index, and their extensions (Congalton and Mead, 1983; Rosenfield and Fitzpatrick-Lins, 1986; Monserud and Leemans, 1992; Pontius, 2000, 2002; Fritz and See, 2005). These measures are often used in remote sensing to compare actual land cover type by location with that estimated from satellite images. However, they are applicable to any tessellations defined by the same or comparable variables.

Visual analysis and quantitative measures are also effective for analyzing tessellations of different attribute types. Although the measures mentioned above are not directly applicable, other summary statistics are available, such as the average area and perimeter of regions, their variance and standard deviation, the spatial mean of their gravity centers, and so forth. Unfortunately, however, unlike χ^2 and the Kappa index, these statistics do not evaluate the differences in tessellations by location. They are calculated for each tessellation separately and compared among tessellations. This implies that they do not reflect differences in the arrangement of regions. Basic geometrical transformations such as translation, rotation, and reflection do not affect statistics values.

A difference in spatial arrangement is not negligible in spatial analysis. To resolve this problem, this chapter proposes a new method for analyzing spatial tessellations, described in Section 11.2. Tessellations are compared in terms of three different aspects: granularity, structure, and hierarchy. They are described conceptually, and a mathematical definition of quantitative measures is given. A method for visualizing the relationship among spatial tessellations is also proposed. In Section 11.3, the method is applied to the analysis of hypothetical and real datasets to evaluate its advantages and limitations. The latter is the analysis of candidate plans for a new administrative system in Japan called the Doshusei system. Section 11.4 summarizes and discusses the conclusions.

11.2 METHOD

This chapter analyzes a set of spatial tessellations defined in the same region. Each tessellation may represent the distribution of different variables or that of the same variable at different times. Examples include administrative units, postal zones, and school districts. To compare tessellations, we consider only their spatial properties, neglecting their nonspatial attributes.

The method consists of the following steps: (1) the introduction of three concepts of similarity in the comparison of spatial tessellations, (2) a mathematical definition of the concepts, and (3) visualization of the relationships among tessellations. These are described in detail in the following sections.

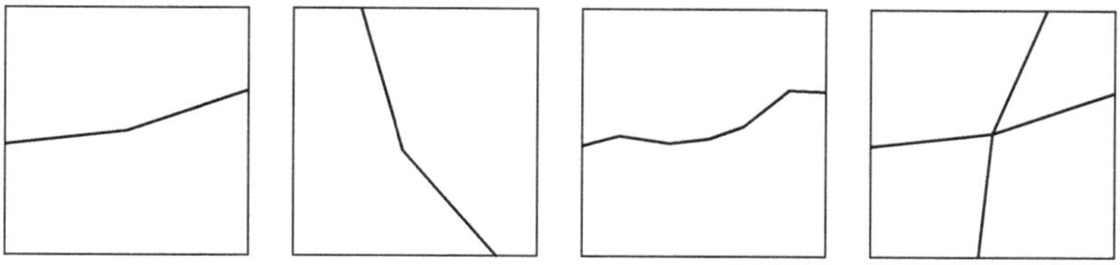

FIGURE 11.1 Three concepts of similarity between tessellations. Tessellations Ω_1 and Ω_2 are similar in spatial granularity but different in the spatial structure of regions. Tessellations Ω_1 and Ω_3 are similar at a global scale in both granularity and structure. Tessellation Ω_4 is obtained by further dividing regions in Ω_1.

11.2.1 Three Concepts of Similarity between Tessellations

This chapter considers three different aspects in evaluating the similarity between tessellations: granularity, structure, and hierarchy. They are described conceptually, after which a formal mathematical definition is given.

Consider the four tessellations shown in Figure 11.1. They are defined over the same square region. The first two tessellations, Ω_1 and Ω_2, look totally different. In terms of granularity, however, they are quite similar because the four regions are almost the same in size. Tessellations Ω_1 and Ω_3 are similar at a global scale in both granularity and spatial structure. Tessellations Ω_1 and Ω_4 are different at least in granularity. Looking at the tessellations in detail, however, we find that the boundary line shared by the two regions in tessellation Ω_1 also appears in tessellation Ω_4. Each region in tessellation Ω_1 is further divided into two smaller regions in tessellation Ω_4.

In the above case, the similarity between tessellations can be described by three spatial concepts: granularity, structure, and hierarchy. Tessellations Ω_1 and Ω_2 are similar in *granularity*, while Ω_1 and Ω_3 are similar in *structure*. Concerning Ω_1 and Ω_4, we can say that the former is a *higher-level tessellation* of the latter, while the latter is a *lower-level tessellation* of the former. This is called the *hierarchical relationship* between tessellations.

11.2.2 Quantitative Evaluation of Similarity between Tessellations

For a formal definition of the three concepts mentioned above, a mathematical framework will be introduced. Consider a set of spatial tessellations defined in region S of area A denoted by $\Theta = \{\Omega_1, \Omega_2, \ldots, \Omega_n\}$. Tessellation Ω_i consists of n_i regions, $\{T_{i1}, T_{i2}, \ldots, T_{ini}\}$. The nature of tessellations ensures the following properties:

$$\bigcap_j T_{ij} = \varnothing\left(i = 1,\ldots,n\right) \tag{11.1}$$

and

$$\bigcup_j T_{ij} = S\left(i = 1,\ldots,n\right) \tag{11.2}$$

One method of representing the tessellations mathematically is to use binary indicator functions:

$$\rho\left(\mathbf{x};T_{ij}\right) = \begin{matrix} 1 & \text{if } \mathbf{x} \in T_{ij} \\ 0 & \text{otherwise} \end{matrix} \tag{11.3}$$

The above representation specifies the region in which every point in S is located. This definition, however, is redundant because the specification of regions is not essential in the definition of a tessellation; a tessellation only determines whether two locations are contained in the same or different regions. We thus use a binary comparison function instead:

$$\begin{aligned} \sigma\left(\mathbf{x},\mathbf{x};\ _i\right) &= \begin{matrix} 1 & \text{if} \sum_j \rho\left(\mathbf{x};T_{ij}\right)\rho\left(\mathbf{x};T_{ij}\right) = 1 \\ 0 & \text{otherwise} \end{matrix} \\ &= \sum_j \rho\left(\mathbf{x};T_{ij}\right)\rho\left(\mathbf{x};T_{ij}\right) \end{aligned} \tag{11.4}$$

This function is calculated from $\rho(\mathbf{x};\, T_{ij})$ and $\rho(\mathbf{x}';\, T_{ij})$. It is 1 if two points $\mathbf{x}$ and $\mathbf{x}'$ are contained in the same region in Ω_i.

Using the function, we evaluate the similarity between tessellations for a given pair of points. Four indicator functions are defined as follows:

$$\begin{aligned} \eta_{11}\left(\mathbf{x},\mathbf{x};\ _i,\ _j\right) &= \sigma\left(\mathbf{x},\mathbf{x};\ _i\right)\sigma\left(\mathbf{x},\mathbf{x};\ _j\right) \\ \eta_{10}\left(\mathbf{x},\mathbf{x};\ _i,\ _j\right) &= \sigma\left(\mathbf{x},\mathbf{x};\ _i\right)\left\{1-\sigma\left(\mathbf{x},\mathbf{x};\ _j\right)\right\} \\ \eta_{01}\left(\mathbf{x},\mathbf{x};\ _i,\ _j\right) &= \left\{1-\sigma\left(\mathbf{x},\mathbf{x};\ _j\right)\right\}\sigma\left(\mathbf{x},\mathbf{x};\ _j\right) \\ \eta_{00}\left(\mathbf{x},\mathbf{x};\ _i,\ _j\right) &= \left\{1-\sigma\left(\mathbf{x},\mathbf{x};\ _j\right)\right\}\left\{1-\sigma\left(\mathbf{x},\mathbf{x};\ _j\right)\right\} \end{aligned} \tag{11.5}$$

Obviously, the following equation holds for any pair of $\mathbf{x}$ and $\mathbf{x}'$.

$$\eta_{11}\left(\mathbf{x},\mathbf{x};\ _i,\ _j\right)+\eta_{10}\left(\mathbf{x},\mathbf{x};\ _i,\ _j\right)+\eta_{01}\left(\mathbf{x},\mathbf{x};\ _i,\ _j\right)+\eta_{00}\left(\mathbf{x},\mathbf{x};\ _i,\ _j\right)=1 \tag{11.6}$$

Integrating the measures above with respect to $\mathbf{x}'$, we obtain

$$
\begin{aligned}
I_{11}\left(\mathbf{x};\ {}_{i},\ {}_{j}\right) &= \int_{\mathbf{x}\in S} \eta_{11}\left(\mathbf{x},\mathbf{x};\ {}_{i},\ {}_{j}\right)d\mathbf{x} \\
I_{10}\left(\mathbf{x};\ {}_{i},\ {}_{j}\right) &= \int_{\mathbf{x}\in S} \eta_{10}\left(\mathbf{x},\mathbf{x};\ {}_{i},\ {}_{j}\right)d\mathbf{x} \\
I_{01}\left(\mathbf{x};\ {}_{i},\ {}_{j}\right) &= \int_{\mathbf{x}\in S} \eta_{01}\left(\mathbf{x},\mathbf{x};\ {}_{i},\ {}_{j}\right)d\mathbf{x} \\
I_{00}\left(\mathbf{x};\ {}_{i},\ {}_{j}\right) &= \int_{\mathbf{x}\in S} \eta_{00}\left(\mathbf{x},\mathbf{x};\ {}_{i},\ {}_{j}\right)d\mathbf{x}
\end{aligned}
\qquad (11.7)
$$

The first equation gives the area of a region in which a point is contained in the same region as **x** in both Ω_i and Ω_j (the light gray area in Figure 11.2). The second equation indicates the area of a region in which a location is contained in the same region as **x** in Ω_i but in a different region in Ω_j (the medium gray area in Figure 11.2). From Equation (11.6) we also have

$$I_{11}\left(\mathbf{x};\ {}_{i},\ {}_{j}\right) + I_{10}\left(\mathbf{x};\ {}_{i},\ {}_{j}\right) + I_{01}\left(\mathbf{x};\ {}_{i},\ {}_{j}\right) + I_{00}\left(\mathbf{x};\ {}_{i},\ {}_{j}\right) = A \qquad (11.8)$$

Using Equation (11.7), we evaluate the similarity between tessellations around a given point **x**. The granularity of tessellation Ω_i around **x** is defined by

$$r_g\left(\mathbf{x},\ {}_{i}\right) = \frac{I_{11}\left(\mathbf{x};\ {}_{i},\ {}_{j}\right) + I_{10}\left(\mathbf{x};\ {}_{i},\ {}_{j}\right)}{A} \qquad (11.9)$$

It is the ratio of points in Ω_i contained in the same region as **x** to those in different regions, and consequently, it depends only on Ω_i.

The structural similarity between tessellations Ω_i and Ω_j is measured by

$$r_s\left(\mathbf{x},\ {}_{i},\ {}_{j}\right) = \frac{I_{11}\left(\mathbf{x};\ {}_{i},\ {}_{j}\right) + I_{10}\left(\mathbf{x};\ {}_{i},\ {}_{j}\right)}{A} \qquad (11.10)$$

It is the ratio of points contained in the same region as **x** or in different regions consistently in both Ω_i and Ω_j (recall Figure 11.2). It ranges from 0 to 1. It is 1 if the region in which **x** is contained in Ω_i is identical to that in Ω_j. It decreases toward 0 as the overlap of the two regions becomes smaller.

Hierarchical similarity is evaluated by two measures. One is defined as

$$r_{hu}\left(\mathbf{x},\ {}_{i},\ {}_{j}\right) = \frac{I_{11}\left(\mathbf{x};\ {}_{i},\ {}_{j}\right)}{I_{11}\left(\mathbf{x};\ {}_{i},\ {}_{j}\right) + I_{10}\left(\mathbf{x};\ {}_{i},\ {}_{j}\right)} \qquad (11.11)$$

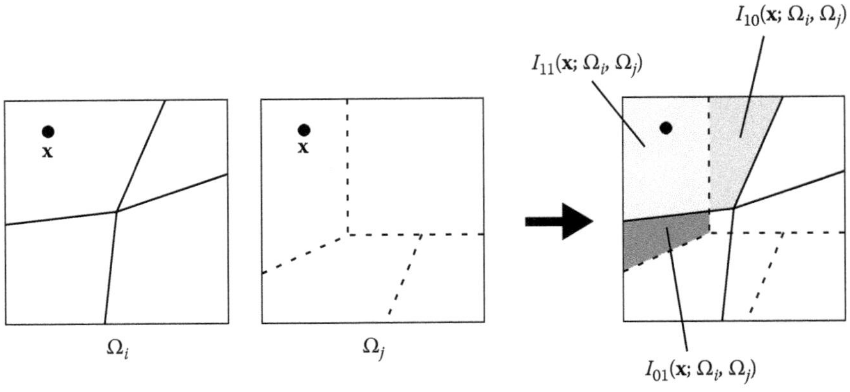

FIGURE 11.2 Evaluation of the similarity between tessellations around a given point.

which indicates the ratio of the points contained in the same region as $\mathbf{x}$ in both Ω_i and Ω_j to the points contained in the same region as $\mathbf{x}$ in Ω_i. It is 1 if the region containing $\mathbf{x}$ in Ω_i fully contains the region containing $\mathbf{x}$ in Ω_j. In such a case, the former region is called a *higher-level polygon* of the latter around $\mathbf{x}$. The measure decreases toward 0 as the hierarchical relationship collapses.

Another measure is given by

$$r_{hl}\left(\mathbf{x}, \ {}_i, \ {}_j\right) = \frac{I_{11}\left(\mathbf{x}; \ {}_i, \ {}_j\right)}{I_{11}\left(\mathbf{x}; \ {}_i, \ {}_j\right) + I_{01}\left(\mathbf{x}; \ {}_i, \ {}_j\right)} \tag{11.12}$$

If the region containing $\mathbf{x}$ in Ω_i is fully contained in the region containing $\mathbf{x}$ in Ω_j, the former region is called a *lower-level polygon* of the latter, and the measure becomes 1. Like $r_{hu}(\mathbf{x}; \Omega_i, \Omega_j)$, $r_{hl}(\mathbf{x}; \Omega_i, \Omega_j)$ ranges from 0 to 1. From the definition of $r_{hl}(\mathbf{x}; \Omega_i, \Omega_j)$, we can easily confirm that

$$r_{hu}\left(\mathbf{x}, \ {}_i, \ {}_j\right) = r_{hl}\left(\mathbf{x}; \ {}_j, \ {}_i\right) \tag{11.13}$$

This implies that it is enough to consider either $r_{hu}(\mathbf{x}; \Omega_i, \Omega_j)$ or $r_{hl}(\mathbf{x}; \Omega_i, \Omega_j)$ in the evaluation of the hierarchy.

From the local measures defined above, we evaluate the degree of similarity at a global scale. The global granularity of tessellation Ω_i is given by integrating $r_g(\mathbf{x}; \Omega_i)$ with respect to $\mathbf{x}$:

$$\begin{aligned} R_g\left(\ {}_i\right) &= \frac{1}{A}\int_{\mathbf{x}\in S} r_g\left(\mathbf{x}; \ {}_i\right)d\mathbf{x} \\ &= \frac{1}{A^2}\int_{\mathbf{x}\in S}\left\{I_{11}\left(\mathbf{x}; \ {}_i, \ {}_j\right) + I_{10}\left(\mathbf{x}; \ {}_i, \ {}_j\right)\right\}d\mathbf{x} \end{aligned} \tag{11.14}$$

This provides a measure of dissimilarity in granularity between Ω_i and Ω_j:

$$R_g\left(\ _i,\ _j\right)=\left|R_g\left(\ _i\right)-R_g\left(\ _j\right)\right| \tag{11.15}$$

Structural and hierarchical similarities are also evaluated by

$$\begin{aligned} R_s\left(\ _i,\ _j\right) &= \frac{1}{A}\int_{\mathbf{x}\in S} r_s\left(\mathbf{x};\ _i,\ _j\right)d\mathbf{x} \\ &= \frac{1}{A^2}\int_{\mathbf{x}\in S}\left\{I_{11}\left(\mathbf{x};\ _i,\ _j\right)+I_{00}\left(\mathbf{x};\ _i,\ _j\right)\right\}d\mathbf{x} \end{aligned} \tag{11.16}$$

$$\begin{aligned} R_{hu}\left(\ _i,\ _j\right) &= \frac{1}{A}\int_{\mathbf{x}\in S} r_{hu}\left(\mathbf{x};\ _i,\ _j\right)d\mathbf{x} \\ &= \frac{1}{A}\int_{\mathbf{x}\in S}\frac{I_{11}\left(\mathbf{x};\ _i,\ _j\right)}{I_{11}\left(\mathbf{x};\ _i,\ _j\right)+I_{10}\left(\mathbf{x};\ _i,\ _j\right)}d\mathbf{x} \end{aligned} \tag{11.17}$$

and

$$\begin{aligned} R_{hl}\left(\ _i,\ _j\right) &= \frac{1}{A}\int_{\mathbf{x}\in S} r_{lu}\left(\mathbf{x};\ _i,\ _j\right)d\mathbf{x} \\ &= \frac{1}{A}\int_{\mathbf{x}\in S}\frac{I_{11}\left(\mathbf{x};\ _i,\ _j\right)}{I_{11}\left(\mathbf{x};\ _i,\ _j\right)+I_{01}\left(\mathbf{x};\ _i,\ _j\right)}d\mathbf{x} \end{aligned} \tag{11.18}$$

All the global measures range from 0 to 1. The structural similarity $R_s(\Omega_i, \Omega_j)$ is equal to 1 only if the two tessellations are identical. When Ω_i is a higher-level tessellation of Ω_j, the hierarchical similarity $R_{hu}(\Omega_i, \Omega_j)$ is equal to 1. When Ω_i is a lower-level tessellation of Ω_j, the hierarchical similarity $R_{hl}(\Omega_i, \Omega_j)$ becomes 1. Small values of the measures imply independency in spatial structure between tessellations.

11.2.3 Visualization of the Relationship among Tessellations

Using the measures, we visualize the relationship among tessellations. Let us consider a matrix $R_g = \{R_g(\Omega_i, \Omega_j)\}$. Since it is a distance matrix, we can classify tessellations by cluster analysis based on R_g. The result is visualized by dendrogram trees, which are very effective for intuitive understanding of the whole structure of similarity among tessellations. It is also possible to use $R_s = \{1 - R_s(\Omega_i, \Omega_j)\}$ instead, which evaluates the dissimilarity of spatial structure among tessellations.

Concerning the hierarchical similarity, the matrices $R_{hu} = \{1 - R_{hu}(\Omega_i, \Omega_j)\}$ and $R_{hl} = \{1 - R_{hl}(\Omega_i, \Omega_j)\}$ are not diagonal, so they cannot be used directly in cluster analysis. To make them diagonal, we compare the elements of matrices with respect to the diagonal line and take the larger ones. Mathematically, this is given by

$$R_h(\ _i,\ _j)=\left\{\max\left(R_{hu}(\ _i,\ _j),R_{hu}(\ _j,\ _i)\right)\right\} \tag{11.19}$$

It is enough to consider either $R_{hu}(\Omega_i, \Omega_j)$ or $R_{hl}(\Omega_i, \Omega_j)$ because they are complementary to each other. Matrix $R_h = \{R_h(\Omega_i, \Omega_j)\}$ serves as a distance matrix in cluster analysis to classify the tessellations.

Another method of visualizing the hierarchical relationship is to focus on its nature as an ordering system. Tessellations are located in a two-dimensional space whose Y-coordinate indicates the granularity measure $R_g(\Omega_i)$. Pairs of tessellations having large $R_{hu}(\Omega_i, \Omega_j)$ values are connected by directed links showing the hierarchical relationship. The threshold of $R_{hu}(\Omega_i, \Omega_j)$ is given arbitrarily in advance. This method generates a network diagram representing the global hierarchical relationship among tessellations.

11.3 APPLICATION

This section applies the method proposed above to hypothetical and real data in order to test the validity of the method. A set of hypothetical tessellations is analyzed first, followed by a real example.

11.3.1 Application to Hypothetical Data

Six hypothetical tessellations defined in the same region are analyzed. As seen in Figure 11.3, every pair of tessellations is different in spatial structure but shares some spatial properties.

We first calculate the global measures to describe the similarity between tessellations. Table 11.1 shows the structural similarity $R_s(\Omega_i, \Omega_j)$ between tessellations. Tessellations Ω_1 and Ω_4, for instance, show a large value ($R_s(\Omega_1, \Omega_4) = 0.90$), which is confirmed by visual impression in Figure 11.3. Large values can also be found for $R_s(\Omega_1, \Omega_2)$ and $R_s(\Omega_2, \Omega_6)$. Although the granularities are different in these cases, the tessellations are similar in global structures.

Hierarchical similarity is shown in Table 11.2. We should note that, unlike R_s, R_{hu} is not symmetrical. The table shows that tessellation Ω_2 is quite similar in hierarchy to Ω_1 and Ω_5 ($R_{hu}(\Omega_1, \Omega_2) = 0.97$, $R_{hu}(\Omega_5, \Omega_2) = 0.96$). This is reasonable because as seen in Figure 11.3, tessellation Ω_2 is obtained by further dividing regions in Ω_1 or Ω_5. Tessellations Ω_5 and Ω_6 also show a large value ($R_{hu}(\Omega_5, \Omega_6) = 0.84$) because Ω_6 is a finer tessellation of Ω_5.

We then apply cluster analysis to the tessellations. Tessellations are classified by the similarity in structure and hierarchy separately. Similarity matrices R_s and R_h are used in the Ward method to obtain dendrograms.

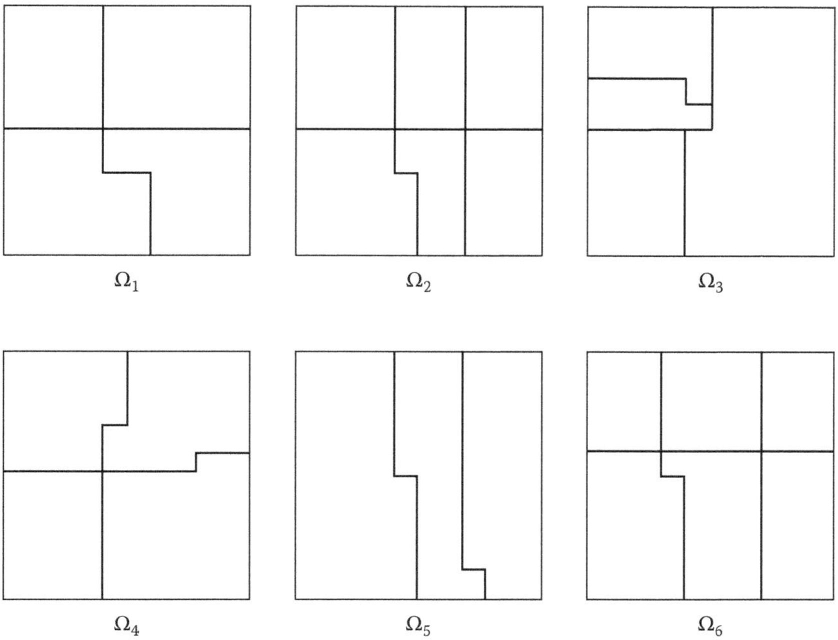

FIGURE 11.3 Six tessellations in the same region.

TABLE 11.1
Structural Similarity $R_s(\Omega_i, \Omega_j)$ between Tessellations

$R_s(\Omega_i, \Omega_j)$	Ω_1	Ω_2	Ω_3	Ω_4	Ω_5	Ω_6
Ω_1	1.00	0.91	0.71	0.90	0.72	0.81
Ω_2	0.91	1.00	0.68	0.87	0.81	0.90
Ω_3	0.71	0.68	1.00	0.78	0.68	0.71
Ω_4	0.90	0.87	0.78	1.00	0.69	0.84
Ω_5	0.72	0.81	0.68	0.69	1.00	0.76
Ω_6	0.81	0.90	0.71	0.84	0.76	1.00

Figure 11.4 shows the dendrogram based on the structural similarity between tessellations. The figure indicates that tessellations Ω_1, Ω_2, Ω_4, and Ω_6 are classified into a single group while Ω_3 and Ω_5 belong to another group of tessellations. For tessellation Ω_3, this is easily confirmed by Table 11.1 and Figure 11.3, because it looks quite different from the other tessellations. On the other hand, tessellation Ω_5 looks similar to Ω_2, at least in Figure 11.3. However, Table 11.1 indicates that Ω_2 is more similar to Ω_1 than to Ω_5, which results in the separation of Ω_5 from the others.

Figure 11.5 shows the dendrogram based on the hierarchical similarity between tessellations. The tessellations are clearly classified into three different groups: {Ω_1,

TABLE 11.2
Hierarchical Similarity $R_{hu}(\Omega_i, \Omega_j)$ between Tessellations

$R_{hu}(\Omega_i, \Omega_j)$	Ω_1	Ω_2	Ω_3	Ω_4	Ω_5	Ω_6
Ω_1	1.00	0.66	0.73	0.80	0.65	0.48
Ω_2	0.97	1.00	0.76	0.86	0.96	0.72
Ω_3	0.45	0.32	1.00	0.54	0.55	0.36
Ω_4	0.80	0.47	0.87	1.00	0.59	0.52
Ω_5	0.47	0.32	0.64	0.43	1.00	0.42
Ω_6	0.69	0.71	0.85	0.77	0.84	1.00

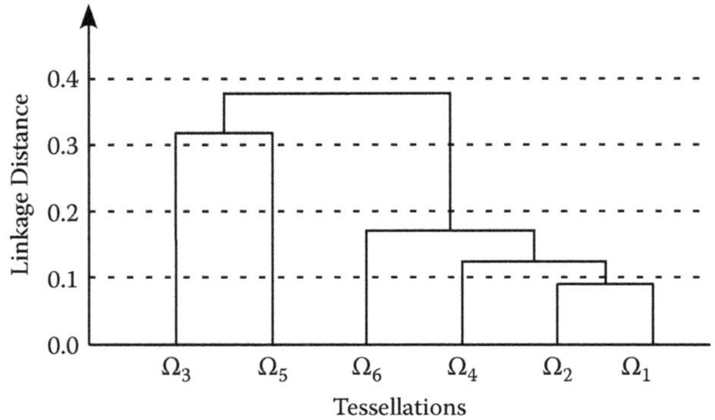

FIGURE 11.4 Dendrogram based on the structural similarity between tessellations.

Ω_2}, {Ω_3, Ω_4}, and {Ω_5, Ω_6}. This is because the hierarchical similarity is more highly evaluated than the structural similarity, which can be confirmed by Figure 11.3.

To consider the hierarchical structure more explicitly, the relationships among tessellations are visualized by a network diagram (Figure 11.6). This representation permits us to understand not only local but also global hierarchical relationships among tessellations. Tessellation Ω_3, for instance, is an upper-level tessellation of Ω_4 and Ω_6, while Ω_2 is a lower-level tessellation of Ω_1, Ω_4, and Ω_5. Tessellations Ω_3, Ω_4, and Ω_2 form a single line of a hierarchical relationship.

11.3.2 Application to a Real Dataset

We then apply the method proposed to the analysis of a set of candidate plans for a new administrative system called the Doshusei system. The Doshusei system is a set of new administrative units now under consideration as part of administrative and financial reforms. Since many plans have been proposed from different viewpoints, it is necessary to classify the plans to find representative patterns of administrative systems and to understand the differences between the plans.

To understand the spatial properties of the plans in greater depth, other spatial tessellations from Japan are also considered. The coverage areas of administrative

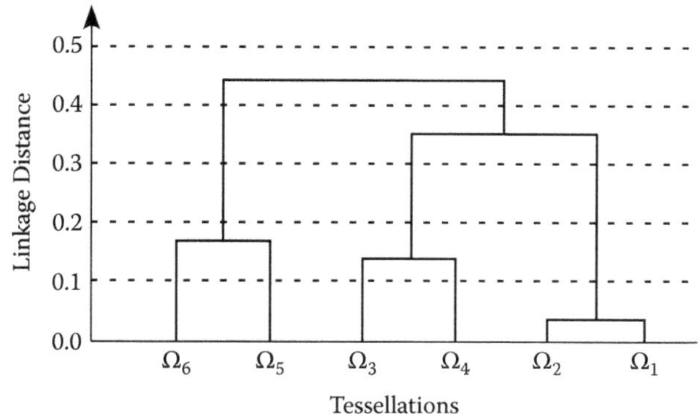

FIGURE 11.5 Dendrogram based on the hierarchical similarity between tessellations.

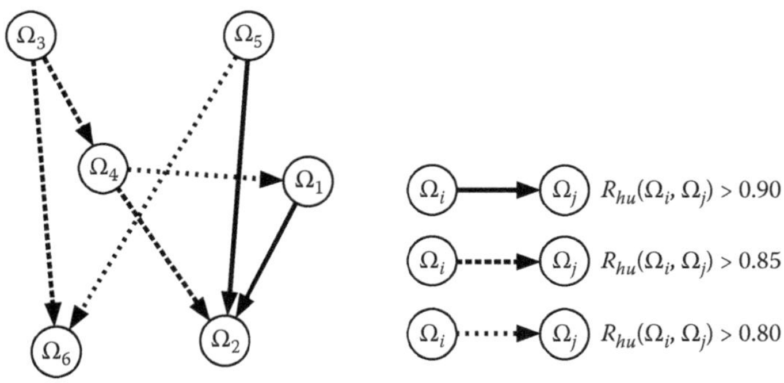

FIGURE 11.6 Network diagram representing the hierarchical relationship among tessellations. The vertical axis is the granularity of tessellations; upper-level tessellations are located higher than lower-level ones. Directed links indicate the hierarchical relationship between tessellations. Different arrow symbols show the closeness of the hierarchical relationship.

branches, nonpublic sectors, and sets of regions classified by socioeconomic and geographical properties are compared with Doshusei plans. As a result, seventy-nine spatial tessellations, all of which are determined based on prefecture units and which cover the whole area of Japan, are analyzed. Every region of tessellations consists of one or more prefectures, and every prefecture is covered by a single region. Figure 11.7 shows an example of a Doshusei system.

Let us first examine the spatial variation in the structural similarity of tessellations. To this end, we evaluate the local structural similarity using

$$\begin{aligned} r_{ls}(\mathbf{x}) &= \sum_i \sum_j r_s(\mathbf{x};\ \ _i,\ \ _j) \\ &= \frac{1}{A} \sum_i \sum_j \left\{ I_{11}(\mathbf{x};\ \ _i,\ \ _j) + I_{00}(\mathbf{x};\ \ _i,\ \ _j) \right\} \end{aligned} \tag{11.20}$$

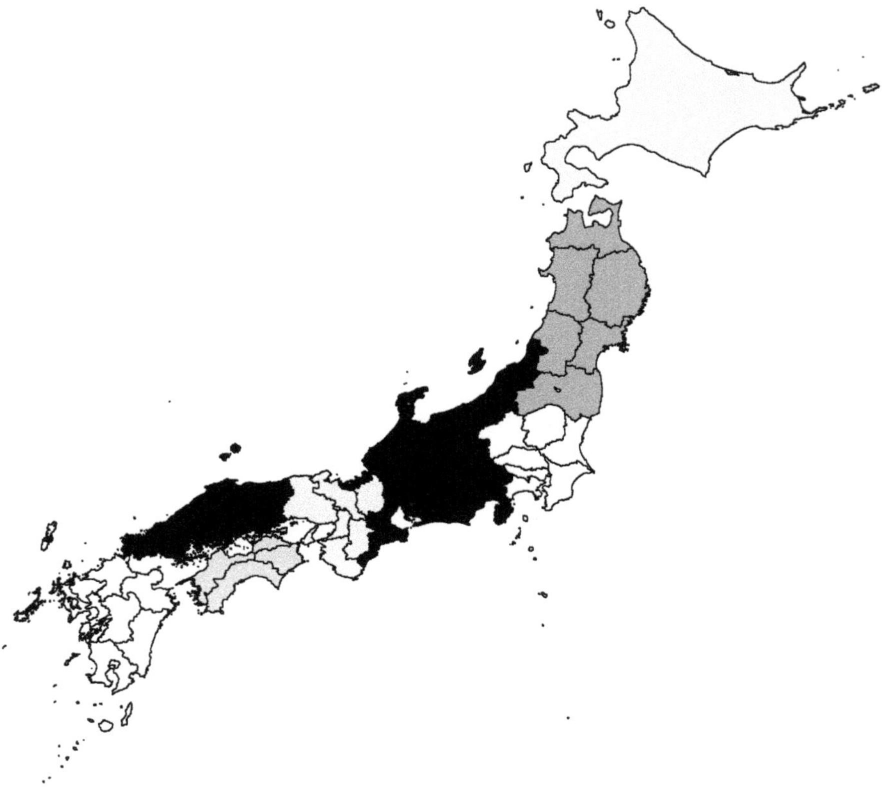

FIGURE 11.7 A Doshusei system. Gray shades indicate Doshusei regions. Black lines indicate the boundaries of prefectures.

Figure 11.8 shows the spatial distribution of the local measure of structural similarity. This figure shows that tessellations are not consistent in the central area of Japan. This is reasonable because prefectures in this area are surrounded by many neighbors, so that numerous combinations are possible to form larger spatial units. In addition, since these prefectures are relatively large in size and population, a local modification of a tessellation system inevitably causes a large effect on the whole system. Prefectures along the northern coastline show large values because there is no common agreement on the boundaries between east and west along the coast.

We then classify all the spatial tessellations in terms of structural similarity by using the Ward method. Figure 11.9 shows that the tessellations are roughly classified into two groups and further classified into six subgroups. Having examined each subgroup in detail, we found several properties that characterize the subgroups (Figure 11.10). Among the six subgroups, four consist of only Doshusei plans while the other two contain both Doshusei plans and other tessellations, such as the coverage areas of administrative branches. Doshusei groups are mainly characterized by the size of Doshusei regions. D-coarse group has larger regions, while D-fine group has smaller ones. Doshusei groups of moderate region size are distinguished by the spatial configuration of regions. In D-moderate A group, prefectures along the northern

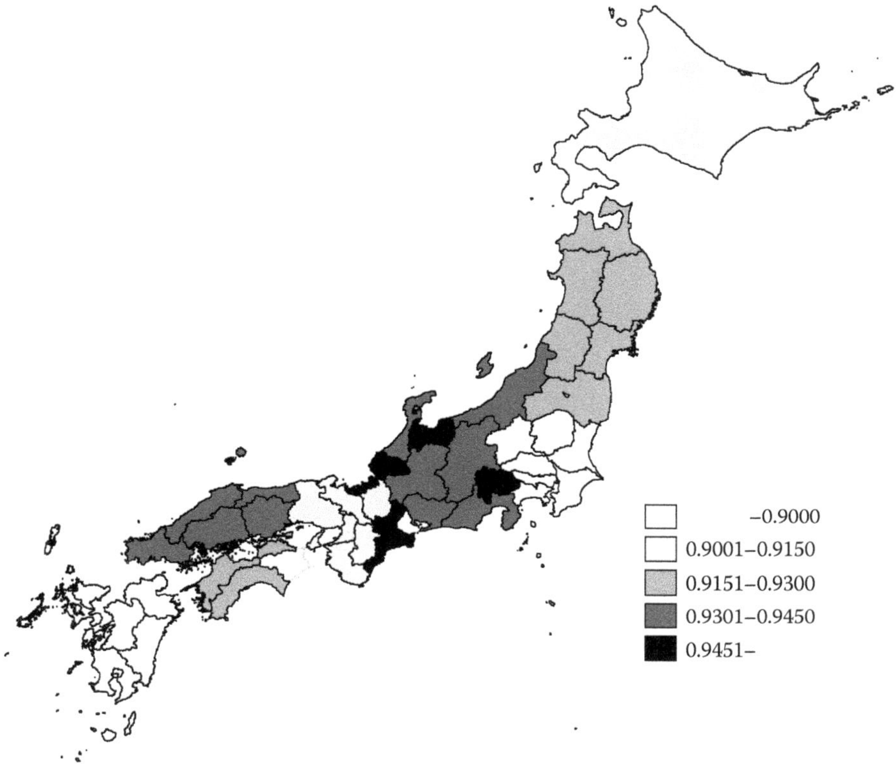

FIGURE 11.8 Spatial variation of structural similarity among tessellations. Large values indicate low consistency in spatial structure among tessellations.

coastline of central Japan are classified into the same region, while D-moderate B divides those prefectures into east and west. Mix A group mainly consists of tessellations of large regions in both Doshusei plans and other tessellations. As shown in Figure 11.10, Mix A and D-coarse groups are considered different from other groups because both mainly consist of tessellations of large regions. Mix B group, on the other hand, consists of tessellations of smaller regions with a wider variation in size.

Looking at the tessellations in each subgroup in further detail, we obtained the following findings:

1. Doshusei plans of moderate region size can be further classified by the spatial configuration of regions in the central area of Japan. D-moderate A group shows two patterns while D-moderate B has three patterns. This is consistent with the result shown in Figure 11.8; the spatial division of the central area of Japan characterizes Doshusei plans in these groups.
2. Doshusei plans are more similar to regional classifications based on socioeconomic properties than to those based on geographical and historical properties.

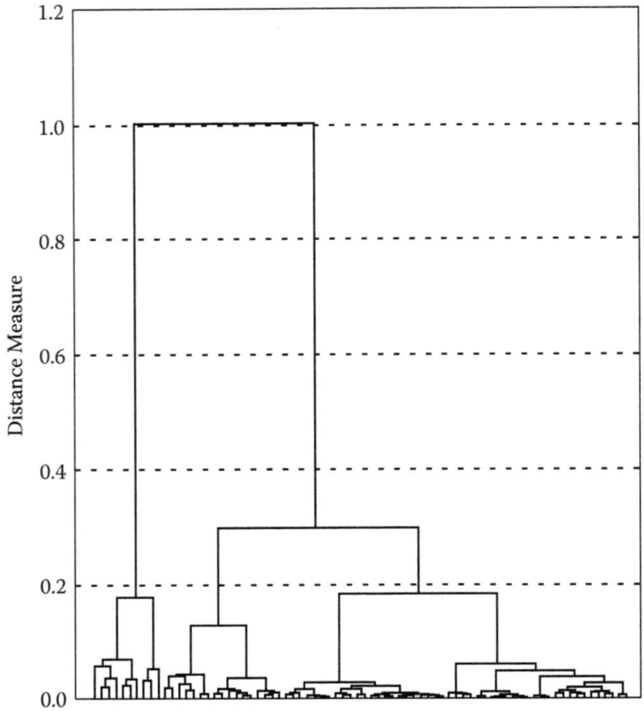

FIGURE 11.9 Classification of 79 spatial tessellations based on structural similarity.

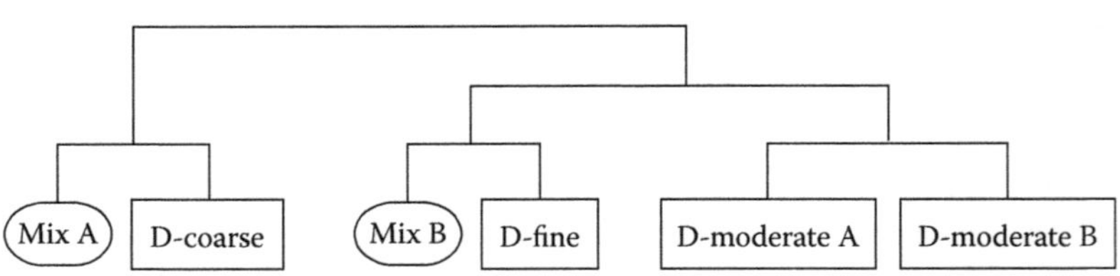

FIGURE 11.10 Six groups of tessellations differing in spatial structure. Groups represented by rectangles consist of only Doshusei plans, while ellipses contain Doshusei plans and other spatial tessellations.

3. Variation in region size is less in Doshusei plans than in other tessellations. One reason for this is that it is more important in administrative systems to keep the spatial units homogeneous in size, population, economic power, and so forth.

The above findings are useful for understanding the relationships among tessellations, such as the similarities and differences between various aspects of tessellations. Such analysis is effective, especially at an early stage of exploratory spatial analysis.

11.4 CONCLUSION

This chapter has proposed a new method for analyzing the similarity between spatial tessellations. It evaluates three aspects of the similarity between tessellations: granularity, structure, and hierarchy. They are evaluated using quantitative measures, which are used in cluster analysis and visualization of the relationships among tessellations. The method was first applied to the analysis of a small set of hypothetical spatial tessellations, in order to deepen our understanding of the properties of measures. This was followed by an application of the analysis to a set of candidate plans for Doshusei systems. Relationships among 79 spatial tessellations, including 33 Doshusei plans, were analyzed by cluster analysis based on structural similarity. This application reveals the properties of the method and its measures as well as empirical findings.

An advantage of the method proposed is that the similarities between tessellations can be evaluated separately in respect to three different aspects. This enables us to avoid subjective interpretation of similarity measures, which often occurs in principal component analysis and multidimensional scaling. Another advantage is that the measures visualize the spatial distribution of similarities among tessellations. As seen in the empirical study, visual representation is effective for understanding the spatial variation in similarity distribution and discussing its underlying structure. It is useful especially for detecting areas in which a significant variation exists in spatial tessellations.

Finally we discuss some limitations of the chapter for future research. First, this chapter focuses on three properties of spatial tessellations in the evaluation of similarities between tessellations. This makes it easy to understand the meaning of quantitative measures and to interpret the results of the analysis. At the same time, however, it limits the scope of our analysis to basic properties of tessellations, such as the size of regions and hierarchical relationships. Other properties, such as perimeter and shape indices, should also be incorporated in analyses. Second, the method was applied to only two sets of spatial tessellations. Further application is clearly necessary to verify whether it is applicable to a wider variety of tessellations. Examples include tessellations representing other types of spatial phenomena, tessellations in other places, and tessellations of the same phenomena at different times. It is also useful to apply the method to tessellations representing natural phenomena because, unlike the tessellations discussed in this chapter, they do not usually share common boundaries.

ACKNOWLEDGMENTS

The authors thank Atsuyuki Okabe, Yasushi Asami, and Toru Ishikawa for their valuable comments.

REFERENCES

Congalton, R. G., and Mead, R. A. 1983. A quantitative method to test for consistency and correctness in photointerpretation. *Photogrammetric Engineering and Remote Sensing* 49:69–74.

Fritz, S., and See, L. 2005. Comparison of land cover maps using fuzzy agreement. *International Journal of Geographical Information Science* 19:787–807.

Kraak, M.-J. 2003. *Cartography: Visualization of geospatial data*. Harlow, U.K.: Pearson Education.

Monserud, R. A., and Leemans, R. 1992. Comparing global vegetation maps with kappa statistic. *Ecological Modelling* 62:275–93.

Nielson, G. M., Hagen, H., and Muller, H. 1997. *Scientific visualization: Overview, methodologies, and techniques*. Los Alamitos, CA: IEEE.

Okabe, A., Boots, B., Sugihara, K., and Chiu, S.-N. 2000. *Spatial tessellations—Concepts and applications of Voronoi diagrams*. New York: John Wiley & Sons.

Pontius Jr., R. G. 2000. Quantification error versus location error in comparison of categorical maps. *Photogrammetric Engineering & Remote Sensing* 66:1011–16.

Pontius Jr., R. G. 2002. Statistical methods to partition effects of quantity and location during comparison of categorical maps at multiple resolutions. *Photogrammetric Engineering & Remote Sensing* 68:1041–49.

Rosenfield, G. H., and Fitzpatrick-Lins, K. 1986. A coefficient of agreement as a measure of thematic classification accuracy. *Photogrammetric Engineering & Remote Sensing* 52:223–27.

Sadahiro, Y. 2002. An exploratory method for analyzing a spatial tessellation in relation to a set of other spatial tessellations. *Environment and Planning A* 34:1037–58.

Slocum, T. A., McMaster, R. B., Kessler, F. C., and Howard, H. H. 2004. *Thematic cartography and geographic visualization*. Upper Saddle River, NJ: Prentice-Hall.

Ware, C. 2004. *Information visualization: Perception for design*. San Francisco: Morgan Kaufmann.

Wright, H. 2006. *Introduction to scientific visualization*. New York: Springer.

12 A New Method of Facility Location Using a Genetic Algorithm Based on Co-Evolution Locational Optimization of Facilities by Co-Evolution of Their Locations and User Allocation

Akio Hori and Tohru Yoshikawa

12.1 AIM AND BACKGROUND OF THE PRESENT STUDY

The present study aims to develop a versatile method of facility location by introducing a biological notion, co-evolution, based on a genetic algorithm (GA) (Holland, 1992), which has attracted attention in recent years as a method for solving difficult problems of vast discrete solution space. The background to the study is described in the following two subsections.

12.1.1 STUDY AS OPTIMAL FACILITY LOCATION

To gain understanding of location planning as a model, previous studies of optimal facility location formulated facility location as a combinational optimization problem, and uniquely allocated users to the facilities according to a facility selection model based on the idea of nearest facility selection or a probability model (Cooper, 1963; Okabe and Suzuki, 1992; Tanimura, 1984a, 1984b, 1986; Osawa, 1992). Also, the allocation of users to facilities was often made for a group of users, e.g., users in a district. Previous studies of GA-based facility location optimization, including those of Tamura and colleagues (1994), Aoki and Muraoka (1996), and Takeda (1999), were also in the same situation.

In these studies, users were allocated uniquely according to the conditions of facilities from a facility-based perspective. For versatile modeling of user's facility selection, allocation of individual users to facilities should also be formulated as a combinational optimization problem, and the facility selection model should be introduced only when users select their facilities.

Aoki and Muraoka (1996) discuss the facility location optimization problem from the user's viewpoint. Takeda (1999) solves the location optimization problem of stopping facilities where users stop by on their way to a destination using the evolution strategy, a kind of GA, and points out the necessity of location and allocation models that are disaggregated according to the user's attribution. To take account of their suggestions, we need to simultaneously formulate the location problem and individual user's allocation problem as a combinational optimization problem to establish a versatile location-allocation problem of facilities from the viewpoints of both facilities and users.

In the present study, we take the suggestions into consideration and propose a new method of solving the facility location optimization problem by using GA based not only on the evolution of facility location, which has been used in previous studies (Tamura et al., 1994; Aoki and Muraoka, 1996; Takeda, 1999), but also on the simultaneous evolution of user allocation to facilities. With various attribution information added to facilities and users, the method is highly versatile and can handle location planning of complex-structured facilities such as those with designated capacity or containment hierarchy structure (Tanimura, 1986).

12.1.2 GA Research

GA researches in the field of architecture and city planning in Japan (Tamura et al., 1994;, Aoki and Muraoka, 1996; Takeda, 1999; Ohsaki, 1998; Takizawa et al., 1998; Hidaka and Asami, 1998; Yamashita and Tomokiyo, 1998; Minemasa et al., 1998; Kawamura, 2000) tend to focus on obtaining optimal results, and few investigate the process of obtaining the results or thoroughly discuss the method applied. GA was originally developed in the field of "artificial life," where its role is to pursue the question of the fundamental nature of life. Therefore, when GA is applied to facility location optimization problems, there may be great significance in not only obtaining the solution but also exploring the essence of the problem in the process of obtaining the solution.

One of the characteristics of the present study as GA research is the simultaneous evolution of two elements of facility location optimization problems, i.e., location of facilities and allocation of individual users, influencing each other. When the location of facilities evolves, the allocation of individual users evolves in accordance with it, which then facilitates the evolution of the location of facilities. This is the notion of so-called co-evolution[1] (Kawata, 1989; Takabayashi, 1995; Ridley, 1993; Eldredge, 1995; Kauffman, 1995) in biology.

12.2 SETTING OF GA IN THE PRESENT STUDY

First, we overview the process of solving the facility location optimization problem by using GA and clarify the meaning of the terms used below.

Each facility location plan, which is a possible solution of a problem, is called an individual. Individuals are processed by a computer and thus need to be represented numerically. The numerical expression of individuals, each number of the expression, and group sequences of the numbers are called, respectively, coding, a gene, and chromosomes. The evaluation index, i.e., evaluation function or target function, which measures the adequacy of the facility location plans, is called fitness.

The optimal solution in GA is derived as follows. First, several individuals (location plan) are prepared. From these individuals, many next-generation individuals are produced through processing called genetic operation, details of which are explained below. Only the individuals with higher fitness survive, and the others are discarded. The same steps of producing next generations are repeated to obtain a better location plan.

Details of the flow of the coding, evaluation function (fitness), genetic operation, and evolution are shown below.

12.2.1 Coding

As stated above, a location plan consists of two elements, location of facilities and allocation of individual users. Therefore, coding uses two chromosomes to present an individual: one for facility location and the other for user allocation. For example, suppose that we have a location plan where two facilities are located in a 4-by-4 mesh area and users form a utility area, as in the upper figure of Figure 12.1. Transforming this plan to an individual in GA, we have the lower figure of Figure 12.1. The first chromosome represents the allocation of individual users where the genes are 0 when individual users use facility 1 and 1 when the users use facility 2. The second chromosome shows the location of facilities, presenting the mesh number of the facility location in binary number representation. These two chromosomes form a single individual, evolving through a co-evolution process.

In the example of Figure 12.1, 40 bits are necessary for each individual, although the number of necessary bits changes depending on the problem that we want to solve. The present method has the disadvantage that it is more difficult to seek a solution than with conventional methods since it has to handle a larger number of bits, and thus the solution space is significantly large, of the order of 2^N when the number of bits is N. While the conventional methods, however, uniquely determine the allocation of users, the present method uses coding for users and enhances the versatility, so this is a trade-off. The versatility can be obtained only when we use the GA, which is a highly capable tool for finding a solution.

12.2.2 Evaluation Function

GA is used to adapt the individuals, introduced in Section 12.2.1, to the environment. The degree of adaptation is called the fitness, and each individual has its own fitness. Fitness is calculated from the evaluation function that is set by the planner. The evaluation function controls the environment. The quality of a model is therefore determined by how much the evaluation function reflects the planner's intention. For example, we could formulate a problem of distance between facilities

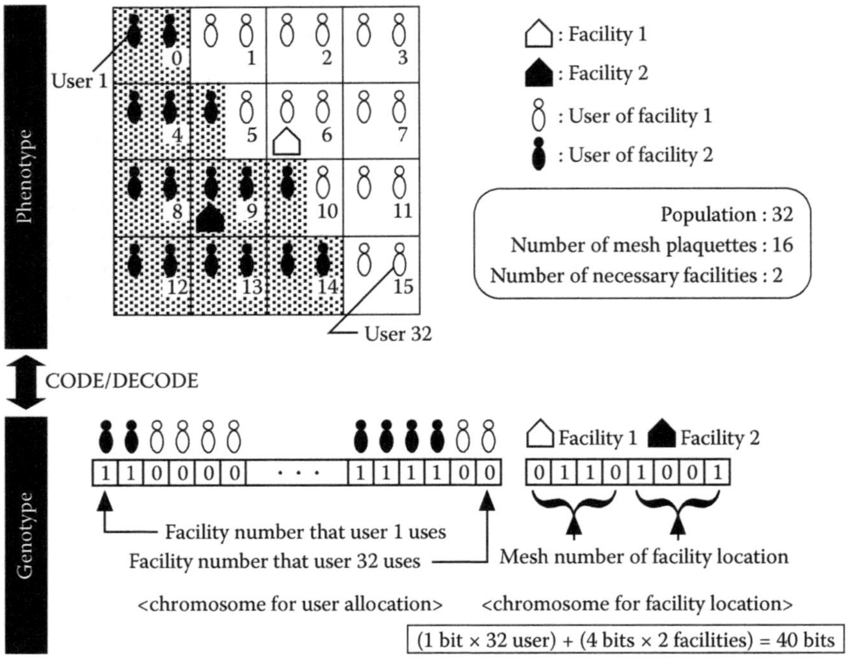

FIGURE 12.1 Proposed coding method.

and users, e.g., mini-sum problem or mini-max problem, by adding information on the position in the mesh space to each gene. We could easily formulate problems of facilities with complicated conditions, e.g., designated capacity or containment hierarchy structure, by introducing various attributions such as sex, age, economic power, or body condition to each gene. This is the most important characteristic of the present method.

12.2.3 Genetic Operation

The genetic operation is important for efficient evolution of individuals while saving the schema, i.e., genetic cluster that enhances the fitness, in the chromosome. In the present study we employed crossover and mutation operations.

The crossover operation produces children from two parent individuals selected under a certain condition in order to transfer the schema to the next generation. In the present study, the crossover position of the first or second chromosome is determined randomly, and the gene clusters of the two parents are exchanged to generate two children, as shown in the left of Figure 12.2.

The mutation operation means randomly replacing genes in chromosomes of individuals with alleles (i.e., replacing 1 with 0 or 0 with 1) to maintain the diversity of species. In the present study, we randomly select a gene position in the first or second chromosome and replace the gene with an allele, as presented in the right of Figure 12.2.

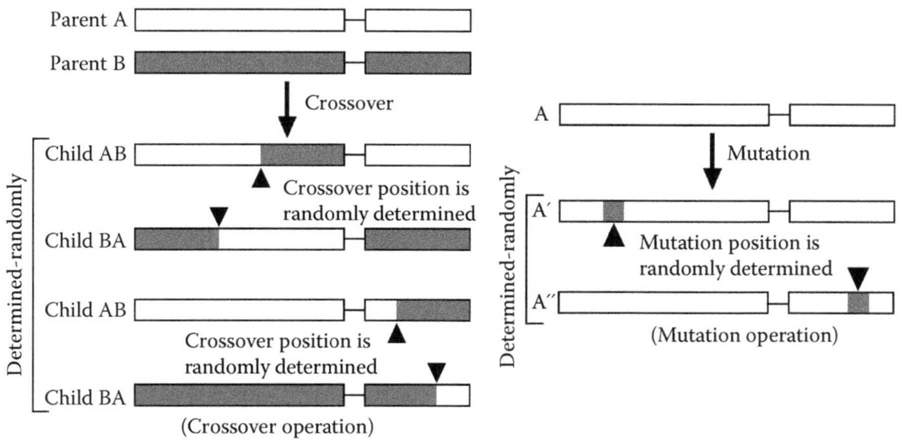

FIGURE 12.2 Crossover and mutation operations.

In both the crossover and mutation operations, only one of the first and second chromosomes is processed. The first chromosome corresponds to the allocation of users and the second to the location of facilities, and the operations change either of them. With a change in one chromosome, individuals adapt to the environment. Then, in accordance with the changed chromosome, the other chromosome changes to adapt to the environment. The operations therefore aim to obtain the effects of cooperation and coexistence through the co-evolution of the two chromosomes.

12.2.4 Flow of Evolution

From the genetic operation explained above, individuals evolved as shown in Figure 12.3. The genetic operation is conducted for all generations, and the selection of parents, except the first-rank parents, is determined by roulette selection[2] with a fitness-related weight. The evolution stops when the generation reaches the designated final generation or when the fitness of the first-rank individual does not change over a designated range of generations. The final generation is called the converged generation, and the first-rank individual of the generation is regarded as the optimal solution. The individuals of the first generation are all generated by random numbers.

The dead individuals in the figure present the individuals that cannot be used as a facility location plan (i.e., when multiple facilities are located in the same mesh plaquette) and will be replaced by individuals of higher fitness.

12.3 APPLICATION TO VIRTUAL CITY

For detailed analysis of the nature and behavior of the present method, we try to optimize the facility location in a virtual city of mesh space in Figure 12.4, by considering a mini-sum problem of minimizing the total moving distance. In particular, we optimize the location of facilities with the designated capacity shown in Section 12.3.1 and with the containment hierarchy structure shown in Section 12.3.2 as sample problems

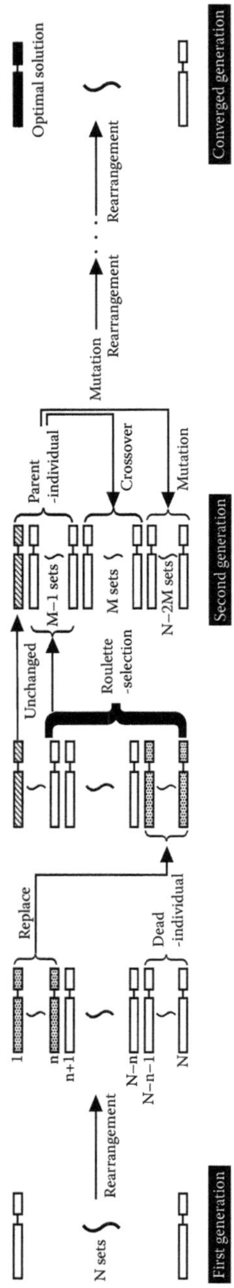

FIGURE 12.3 Flow of evolution.

0	1	2	3	4	5	6	7
8	9	10	11	12	13	14	15
16	17	18	19	20	21	22	23
24	25	26	27	28	29	30	31
32	33	34	35	36	37	38	39
40	41	42	43	44	45	46	47
48	49	50	51	52	53	54	55
56	57	58	59	60	61	62	63

8 × 8 = 64 mesh

FIGURE 12.4 Virtual city.

that can be formulated easily with the present method. In Section 12.3.3, we analyze the behavior of two chromosomes using the optimized result shown in Section 12.3.1 in the convergence of the evolution, and constitutively interpret the relation between the evolution process in GA and the development process of the location plan.

12.3.1 Optimal Location of Facility with Designated Capacity Condition

12.3.1.1 Prerequisite

We consider the location problem of four facilities in a mesh space with a total of 64 users distributed uniformly. Positional information is given to each gene. Let Z denote the evaluation function. The function is given by the equation

$$Z = \sum_{i=1}^{m,} d(i) \bullet \alpha + \sum_{j=1}^{n} \frac{\left(pf(j) - capa(j)\right)}{n} \tag{12.1}$$

where $d(i)$ denotes the moving distance of user i; $pf(j)$ denotes the number of users at facility j; $capa(j)$ denotes the capacity of facility j; m denotes the total number of users; n denotes the total number of facilities; and α and β are parameters. The optimal plan is given by minimizing Z.

The first term of the evaluation function presents the total moving distance of users in unit of mesh plaquette. The second term of the evaluation function is the square-mean value of the deviation of the actual number of users from the capacity as a penalty for the deviation. We add the second term in order to take account of the facility's capacity. The facility's capacity is an important factor for both the facilities and users in maintaining a certain service level. We use the square value since the service level is dramatically lowered if the deviation of the number of users from the capacity becomes large. The facility's capacity is set to 16 people.

To adjust the unit of the two terms, weight control parameters α and β are multiplied to the first and second term, respectively.[3] Since it is difficult to determine these

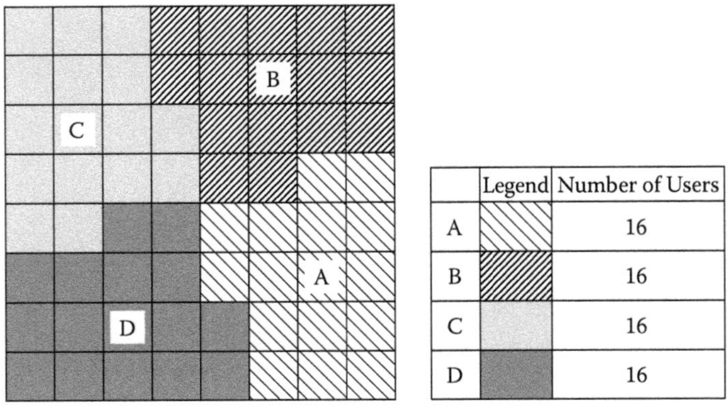

FIGURE 12.5 Optimization result (facility location, utility area, and number of users).

parameter values, we conducted tests using the parameter sets (α, β) = (10,1), (5,1), (1,1), (1,100) and employed (10,1), which gave the most optimal solution. With this parameter setting, a penalty of 0.05 is imposed on the total moving distance if one of the four facilities has the users of the capacity plus one and another facility has the users of the capacity minus one.

12.3.1.2 Results and Discussion

The optimization result is shown in Table 12.1[4] and Figure 12.5. The facility location and utility area (Figure 12.5) were perfectly point-symmetric, and no variation was found in the number of users at each facility, which is 16, owing to the restriction of the capacity. We can conclude that this is the complete solution as it coincides with the optimal solution obtained by the complete count survey under a certain condition.[5] The complete solution was derived through only 1.67 million searches from the solution space, where the total number of combinations was 5.7×10^{45}. This shows the distinctive solution-finding capability of GA.

Interestingly, the utility area has a complicated shape that is rarely obtained intuitively.

12.3.2 Optimal Location of Facilities with Containment Hierarchy Structure

12.3.2.1 Review of Previous Studies

Studies of optimal location of facilities with containment hierarchy structure are classified in two types: one type includes studies by Suzuki (1990), Okunuki and Okabe (1995), and Kishimoto (2000), which theoretically discuss the location optimization problem in a continuum space, and the other type is the study by Nosaka and Yoshikawa (1999), which empirically optimizes facility location in a discrete space. In continuum spaces, theoretical optimization is feasible but empirical optimization is difficult. In discrete spaces, on the other hand, empirical case study is feasible, but the total number of combinations is so huge that we have to make phased optimization in each hierarchy, limiting the solution space. These two types thus compensate for each other's advantages and disadvantages.

TABLE 12.1
Optimization Result

Fitness					
Initial Generation	Converged Generation	Number of Converged Generations	Number of Searches	Total Number of Combinations	Computation Time
2,214.81	984.05	5,580	1,674,000	5.7×10^{45}	6: 01.22

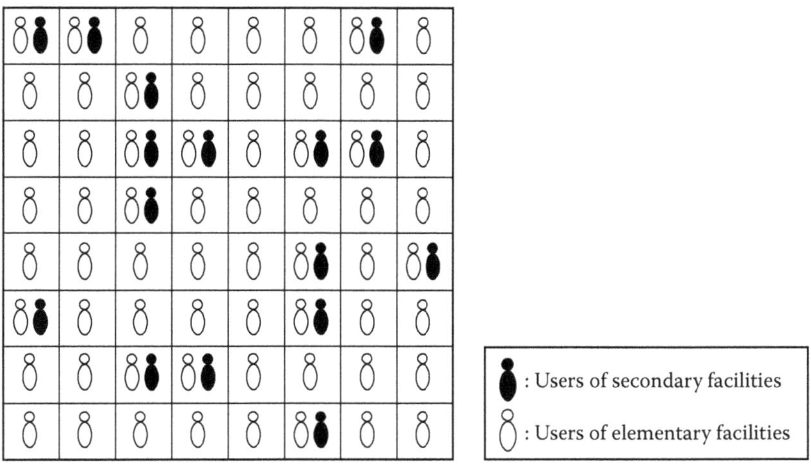

FIGURE 12.6 Distribution of users (of secondary and elementary facilities).

Although the present method belongs to the latter type, optimization with a hierarchy structure in discrete spaces can be performed by adding the hierarchy information to genes.

12.3.2.2 Prerequisite

We consider the location problem of four facilities (two secondary and two elementary facilities) with 16 users of the secondary facilities and 64 users of the elementary facilities distributed over the mesh in Figure 12.6. Positional and hierarchy information is added to each gene. As in Section 12.3.1, the evaluation function, denoted by Z, consists of a term representing total moving distance and a term representing the deviation of the actual number of users from the capacity of eight people in each secondary facility and sixteen in each elementary facility for each generation. The evaluation function is given by

$$Z = \sum_{i=1}^{m_S} d_S(i)\cdot\alpha + \sum_{i=1}^{m_E} d_E(i)\cdot\beta + \sum_{j=1}^{n_S} \frac{(pf_S(j) - capa_S(j))^2}{n_S} \cdot \tag{12.2}$$

$$\gamma + \sum_{j=1}^{n_E} \frac{(pf_E(j) - capaE(j))^2}{n_E} \cdot\delta$$

Here the symbols used in the equation are defined by:

$D_S(i)$, $d_E(i)$: Moving distance of user *i* (*S*: secondary facility, *E*: elementary facility).

$pf_S(j)$, $pf_E(j)$: Number of users of facility *j* (*S*: secondary facility, *E*: elementary facility).

m_S, m_E: Total number of users (*S*: secondary facility, *E*: elementary facility).

n_S, n_E: Total number of facilities (*S*: secondary facility, *E*: elementary facility).

$capa_S(j)$, $capa_E(j)$: Capacity of facility *j* (*S*: secondary facility, *E*: elementary facility).

α, β, γ, δ: Parameters. The parameters chosen, as in Section 12.3.1, are $\alpha = \beta = 10$ and $\gamma = \delta = 1$.

12.3.2.3 Results and Discussion

The optimization result is shown in Table 12.2 and Figure 12.7. The utility area (Figure 12.7) was almost completely point-symmetric, and no variation was found in the number of users at each facility.

In comparison of this facility location with the result shown in Section 12.3.1 (Figure 12.5), the elementary facilities C and D are located in the same place, but the secondary facilities A and B migrate by one mesh toward the center because the users of the secondary facilities are distributed over the entire area.

TABLE 12.2
Optimization Result

Fitness		Number of			
Initial Generation	Converged Generation	Converged Generations	Number of Searches	Total Number of Combinations	Computation Time
2,577.51	1,346.6	7,185	2,155,500	2.2×10^{55}	8: 03.66

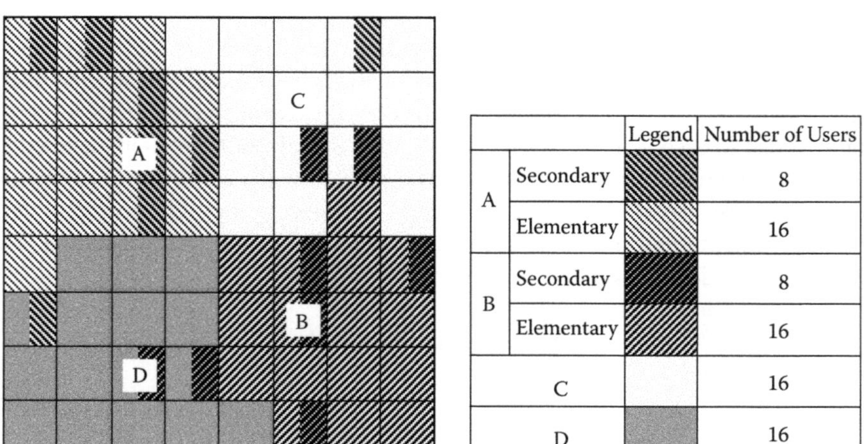

FIGURE 12.7 Optimization result (facility location, utility area, and number of users).

In the user allocation, users of the secondary facilities selected the nearest facilities, while some users of the elementary facilities, for example, those on the leftmost and fourth plaquette from the bottom, did not select the nearest facilities. This is because the locations of the facilities A and B changed from those in Figure 12.5 and multiple facilities appeared nearest to the elementary facility users. An example of this is shown on the leftmost mesh plaquette fourth from the bottom, which is actually assigned to the nearest facility. Since there are multiple choices in determining to which facility the users of these plaquettes are allocated under the restriction of the capacity, there is more than one solution of the same fitness.

Taking these findings together, a solution, not a complete solution though, of the integrated optimization, including the hierarchy structure, can be obtained with the present method.

12.3.3 Analysis of Evolution Process

12.3.3.1 Change of Fitness of Optimum Individual and Fitness Range

The change of the fitness of the optimum individual in Figure 12.8 consists of two parts: rapid evolution and long-term gradual evolution. The former is major evolution, where a variety of individuals (facility location plans) are generated until about the three hundredth generation, and the latter is minor evolution, where the individuals are narrowed down and capable individuals are selected[6] (Eldredge, 1995).

The range of the fitness of each generation in Figure 12.9 shows the diversity of the species in this major evolution. The species of the first generation is extremely diversified and has a large range. However, as the generations progress, the fitness becomes lower and the diversity of the species is reduced and has a small range. In the progress of the generations, we see that the individuals are narrowed down and capable individuals are selected. On the other hand, the fitness range becomes larger in some generations, which is the result of the above-mentioned roulette selection.

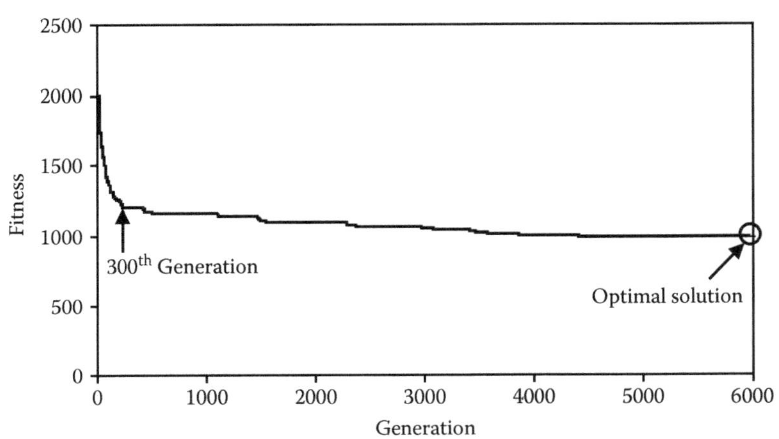

FIGURE 12.8 Change of fitness of the optimum individual.

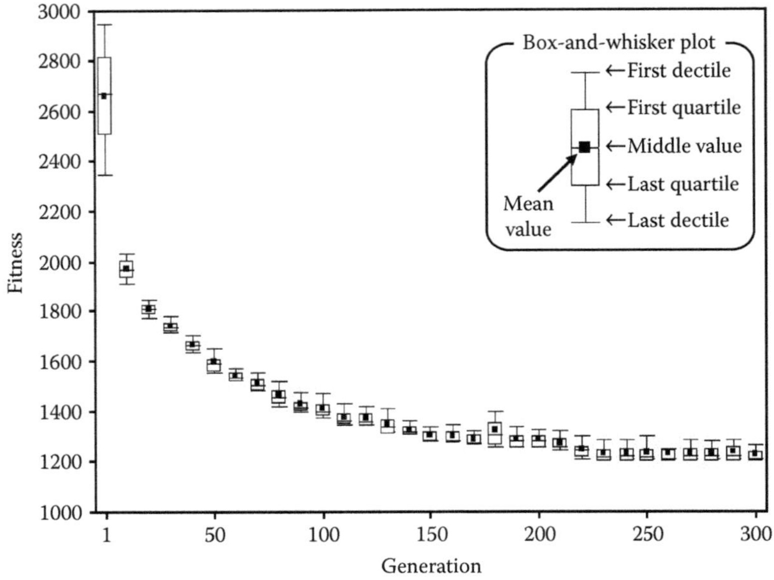

FIGURE 12.9 Change of fitness range of 250 upper-rank individuals.

12.3.3.2 Evolution Process of GA and Development Process of Location Plan

To understand the development process of actual location plans, Figure 12.10 shows part of the evolution process of the optimum individual of each generation. What is important here is which genetic operation generated the individual. This is classified by the choice of the two chromosomes in the figure (the white arrow shows the first chromosome and the black arrow the second one). The arrows change from generation to generation, but their overall behavior is the repetition of the pattern, whereby several generations of the white arrows continue after the appearance of the black arrow. Since the first chromosome represents the user allocation and the latter the facility location, we see that the facility location plans are developed by the repetition of the pattern in which the location is determined first and then the user allocation is changed with the facility locations. This indicates the phenomena of cooperation and coexistence, where the two chromosomes are competing with each other for their improvement, namely, the co-evolution in GA, where the two chromosomes evolved exactly as suggested by the red queen hypothesis.[1] This also resembles the human thinking process: when people think of facility location plans, i.e., the method of learning by trial and error, they improve their plan by repeating the operation whereby either the facility location or the user allocation is changed while the other is fixed.

More detailed analysis revealed that the trial-and-error process in the user allocation could be classified to two patterns, as shown in Figure 12.11. In both patterns, the user allocation eventually evolves. In pattern 1, useless plans of the

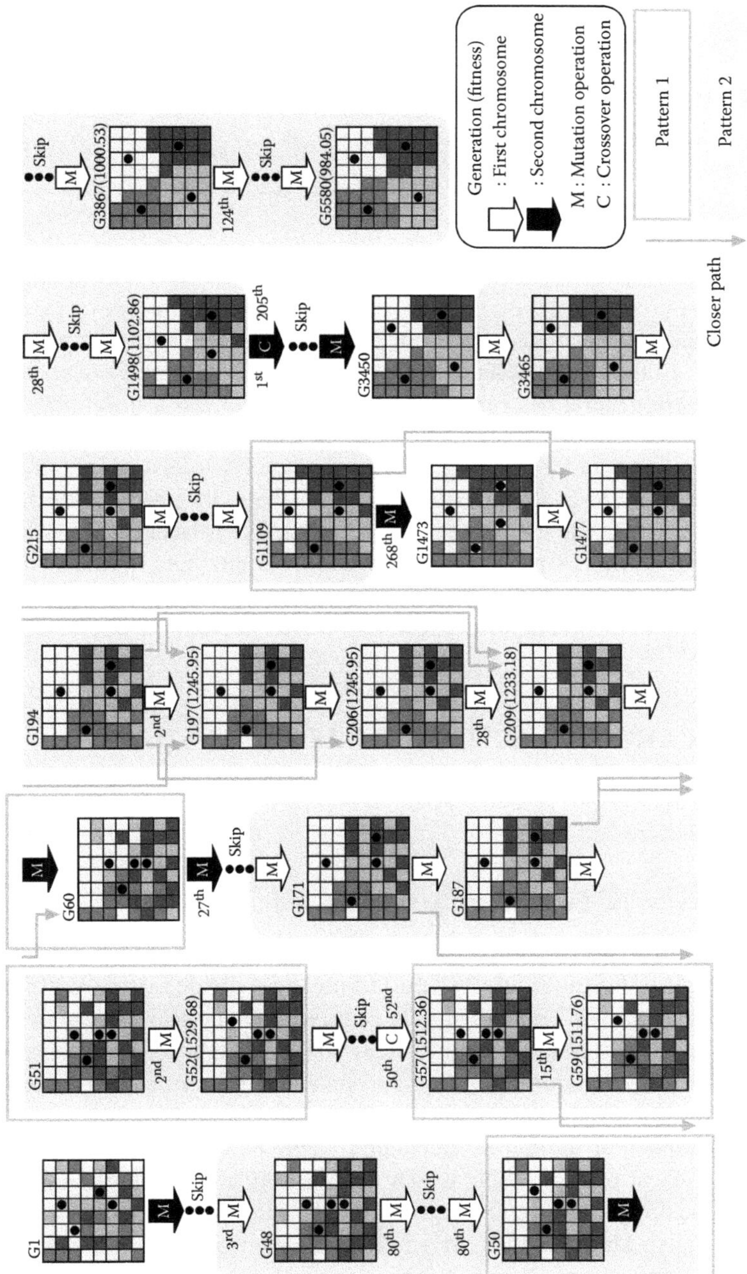

FIGURE 12.10 Evolution process of the optimum individual of each generation.

(Pattern1)

	Generation (n)	→	Generation (n+1) <trial-and-error generation>	→	Generation (n+2)
Allocation of users	A		A		A′
Location of facilities	a		b		a

(Pattern2)

	Generation (n)	→	Generation (n+1) <trial-and-error generation>	→	Generation (n+2)
Allocation of users	A		B		A′
Location of facilities	a		a		a

FIGURE 12.11 Pattern of trial and error.

facility location were generated, while in pattern 2, those of the user allocation were produced. In the final results of the evolution process, there were some useless trial-and-error generations generated even when there was a closer path to the final solution. This evolution seems to take the long way round, but trial and error is intrinsically a learning form to find a better solution regardless of taking a long route to the solution. The evolution process in GA is therefore similar to the human process of thinking. People use intuition based on their experience and knowledge when they think, while GA processes are random. GA attempts to find a solution steadily without being influenced by preconceived ideas or other arbitrary factors. In this regard GA has a solution-finding ability different from the one that people have.

12.4 APPLICATION TO ACTUAL CITY AS A CASE STUDY

In this section, we apply the present method to actual cities as a case study to verify its usefulness.

12.4.1 Overview of Target Area and Prerequisite

Our target area was selected as 6-Chome in the northern area of Karasuyama, Setagaya-ku, Tokyo, Japan (Figure 12.12). The area was divided into 195 (15 × 13) 100 m mesh plaquettes. We optimally located daycare and day-service centers in this area. The centers have a containment hierarchy structure with three services: intensive care (tertiary), ambulatory-type care for demented elderly people (secondary), and preventative care (elementary).

The total number of users was calculated from the estimated population of elderly people (65 or over) in 2020, which was derived by the cohort change rate method, and the appearance ratio of elderly people in different conditions, which was given by Tokyo Metropolitan Government. The number of necessary facilities was calculated from these values and the capacity of the facilities, 45 for tertiary facilities, 16 for secondary, and 90 for elementary, as shown in Table 12.3 (Nosaka and Yoshikawa, 1999). The users in each block area were distributed over the mesh in the same manner[7] as performed by Nosaka and Yoshikawa (1999).

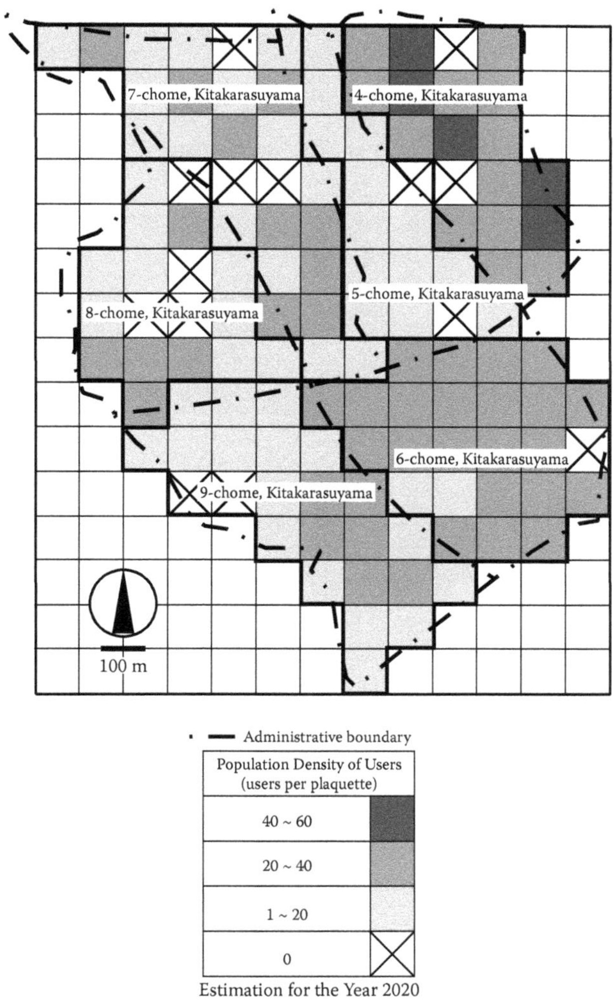

FIGURE 12.12 Target area and population density of users in 2020.

12.4.2 Results and Discussion

The optimization result is shown in Table 12.4 and Figure 12.13. The setting is more complicated than in Section 12.3, and the total number of combinations is huge (3.2 $\times 10^{469}$), although GA allows us to reach a solution through searching about 100 million individuals.

We first discuss the utility area. The result that we obtained showed no separated utility areas, i.e., multiple areas of the same utility appearing across different utility areas, despite the complex setting. In conventional systematic user allocations (Nosaka and Yoshikawa, 1999), since facilities with users up to capacity refuse to accept more users, the presence of the users who cannot select the nearest facilities results in separated areas. Focusing on this problem, Hori and Yoshikawa (1999) employed a flexible user allocation method of allowing a certain range of the facility

TABLE 12.3
Number of Users and Necessary Facilities in 2020

	Population of Elderly People	Upper-Class Facilities	Middle-Class Facilities	Lower-Class Facilities
4-chome, Kitakarasuyama	480	5	6	91
5-chome, Kitakarasuyama	137	0	0	26
6-chome, Kitakarasuyama	622	8	8	118
7-chome, Kitakarasuyama	522	7	7	99
8-chome, Kitakarasuyama	300	5	5	57
9-chome, Kitakarasuyama	327	0	6	62
Total	**2388**	**25**	**32**	**453**
Number of Necessary Facilities	**—**	**1**	**2**	**6**

TABLE 12.4
Optimization Result

Fitness					
Initial Generation	Converged Generation	Number of Converged Generations	Number of Searches	Total Number of Combinations	Computation Time
24,764.5	11,566.07	164,750	98,850,000	3.2×10^{469}	about 10 days

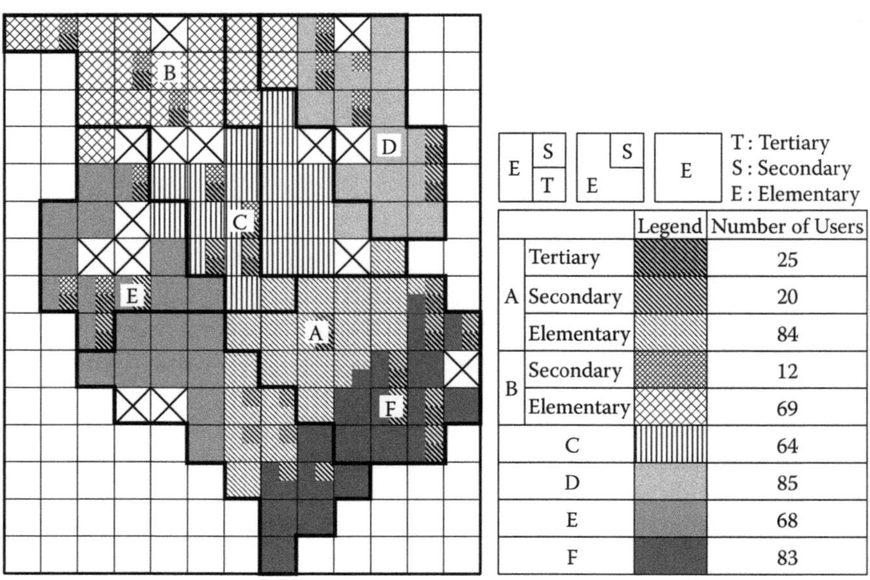

FIGURE 12.13 Optimization result (facility location, utility area, and number of users).

capacity to eliminate separate areas. The method resulted, however, in the variation of the number of the users at each facility, and there was an essential problem of how large a capacity range one should set. On the other hand, the evaluation function in the present method has a term that favors nearest-facility selection and a term that sets the capacity of facilities. The two terms are in a trade-off relation and create a certain kind of equilibrium. Therefore, we obtained a well-balanced solution with no significant variation of the number of users and no separated areas.

Next we discuss the facility location. The tertiary facility, A, is located in a place closer to the southeast region, with a relatively higher population. The secondary facilities, A and B, divide the entire area along a line from northeast to southwest. Many users are allocated in A, which is located in the highly populated southeast area. The elementary facilities, A, B, C, D, E, and F, are relatively unevenly distributed in the southeast area owing to the population density of the users, but mostly spread over the entire area. Thus, the present method exactly optimizes the hierarchy of the facility location, which is significantly affected by user distribution, and introduces a solution that subtly balances the moving distance and capacity of facilities.

From the above, we see that the present method not only solves simultaneously the two problems, the facility location and the user allocation, but also searches for a solution with a delicate balance of multiple elements related to the facility location plan, e.g., user distribution, moving distance, containment hierarchy structure, and capacity of facilities. Therefore, the present method is a practical method for processing in an integrated way the complicated and mixed elements of actual cities.

12.5 SUMMARY

In this chapter we focused on the fact that the optimal facility location problem consists of two issues, facility location and user allocation, and we explicitly used the allocation of individual users. By regarding the facility location and the user allocation as chromosomes that form an individual in GA, we proposed a new method of acquiring an optimized solution by co-evolving the two chromosomes.

We summarize the knowledge that we obtained as follows:

- For the optimal facility location problem, we formulated the location problem of facilities and the allocation problem of individual users simultaneously as a combined optimization problem, and developed a method to solve the problem with GA by co-evolving the location and allocation. As a result, we successfully derived a complete solution to the location optimization problem of the facilities with designated capacity through a short search process in a vast solution space. For the optimal location of the facilities with containment hierarchy structure, we optimized the hierarchical structure in an integrated manner. For the location optimization of the daycare facilities for elderly people, we also reached a solution by making a subtle balance of various elements related to the facility location plans. We consider from these results that our formulation and solution finding method based on GA is versatile and practical as a method for solving facility location optimization problems.

- Analyzing the evolution process of individuals, we found that the fitness changed at two evolution speeds, major evolution and minor evolution. In the evolution of the location plans, the facility locations changed first, and then the user allocation gradually changed at the changed facility locations. They thus co-evolved. The solution search in GA progressed through a trial-and-error process, which is similar to the human thinking process.

We intend to conduct the following research based on our present findings:

- This method aims at a more versatile approach to a variety of problems by setting an evaluation function based on the planner's intentions. In this chapter, we formulated an evaluation function on the basis of the most orthodox mini-sum problem of distance to examine the usefulness of the method. In addition, by surveying user's interests and actual living conditions, we were able to add individual user's sense of value to genes and optimize problems taking account of all the elements in a modern city, which is in fact a complex system. An experimental study would be necessary for this purpose.
- The setting of GA is empirical and difficult to support theoretically. In the present study, we conducted several experiments and finally reached the setting in the virtual city study (Section 12.3). In the case study of an actual city (Section 12.4), we used the setting. However, since the setting in GA depends on the details of the target problem, there should be a setting that is more suitable to the problem and enables a more efficient way of finding the solution. In particular, it is important how we constitute the evaluation function to calculate fitness and determine the parameters used. These technical aspects of GA require a theoretical basis.
- In this study, we composed an individual (facility location plan) of two chromosomes, i.e., facility location and user allocation. As a consequence, subsequent generations inherit both the facility location and the user allocation, and some combinations of the facility location and the user allocation have a higher likelihood of occurrence and some have a lower one. On the other hand, if we regard the facility location and the user allocation as different species and co-evolve them, we can treat them separately. This will allow a higher degree of freedom for the genetic operations to generate subsequent generations. We need to review this approach.
- Other than GA, Cooper's method (Cooper, 1963) is a calculation method of facility location optimization. We explicitly treated the allocation of individual users and used a setting different from that used in conventional methods, and so it is difficult just to compare the present method and Cooper's. Our method, for example, allocates individual users taking account of the capacity of the facilities, while Cooper's method (Cooper, 1963) does not have a limitation in source capacity. When we have a small number of users, however, the difference between our method and Cooper's could lie in their structures, and we may be able to compare them by actually solving problems with both methods. We intend to attempt this comparison in future research.

- For practical application, we divided an actual area into a mesh, which may have the following errors: (1) error of user population density in the allocation of users on the mesh, (2) error of distance in approximating the location of users to the center of each mesh plaquette, (3) error of distance in approximating the location of facilities to the center of each mesh plaquette, and (4) error of distance in approximating road distance to linear distance.

As Nosaka and Yoshikawa (1999) point out, it requires a tremendous amount of time to eliminate the error (1) of the user allocation on the mesh. We can, however, expect future reduction of the error, since it has become possible to process details of population distribution data, such as the base unit area data in the national census, by using geographic information systems (GIS). For (2) and (3), there are studies by Kurita (1993) and Tagashira and Okabe (1998), although more experimental studies are necessary for practical application of the results. Regarding (4), Koshizuka and Kobayashi (1983) point out that the two distances have a certain relationship. Since all the errors should be taken into account when the present method is applied to actual cases, quantitative evaluation study of the errors is necessary.

ACKNOWLEDGMENT

Part of the study was presented orally by the authors (Hori and Yoshikawa, 2000). We are grateful to the audience, M. Nosaka, and anonymous reviewers for their valuable comments. The figures and tables are taken from Hori and Yoshikawa (2001) with permission.

END NOTES

1. *Evolution* generally indicates the adaptation of individual species to the environment. Here, the environment is a physical one such as the landform or climate in which organisms are living. However, as species are also affected by other species in the living environment, which are also evolving, they must constantly evolve. Leigh M. Van Valen called this theory the red queen hypothesis (Ridley, 1993; Eldredge, 1995). Evolution of organisms influencing each other is called co-evolution. In the present study, two chromosomes that constitute an individual in GA are regarded as organisms of different biological species and are co-evolved.
2. This is a method of selecting parent individuals to transfer their genes to the next generation. The selection is made with a probability with a fitness-related weight. The probability of the n-th individual in order of fitness is given by

$$P(n) = \frac{\text{fitness}(n) - \text{fitness}(N)}{\text{fitness}(1) - \text{fitness}(N)} \tag{12.3}$$

where *fitness*(n) denotes the fitness of the n-th individual and N denotes the total number of individuals. A characteristic of the roulette selection is that even lower-rank individuals can become parents, which creates diversified species and activates the evolution, preventing it from falling into a local solution.

3. Only the ratio of α and β is necessary to solve this problem, and thus we need only one of the two parameters. Considering future versatility of the present study, however, we set both parameters to give a meaning to the value of the evaluation function, e.g., to regard the evaluation function as the running cost of the facilities. Since we did not give any meaning to the evaluation function Z in this chapter, α and β, respectively, have the unit of the inverse of distance and the inverse of squared number of users, if we suppose that the evaluation function is dimensionless.
4. The calculation was performed on a Japan IBM workstation (Power2, 66.6 Mhz, 128 MB).
5. The condition is that the solution is obtained in the conventional way, whereby the users select the nearest facilities after the determination of the facility location. In our problem, it can be formulated as a combination optimization problem to select the location of four mesh plaquettes from 64, and thus the total number of combinations is ${}_{64}P_4$. It is, however, too large for a computer to handle, and we made it ${}_{64}C_4$, restricting the solution space.
6. After a long "static" period in which the species did not change much, a large evolution suddenly occurred. The evolution progressed when this process was repeated. The evolution seems to follow the punctuated equilibrium theory (Eldredge and Gould, 1972) in evolutionary biology.
7. Distribution of the users in block areas over the 100 m mesh was conducted with weight 1 for residential areas, 0 for nonresidential areas, and 0.5 for others.

REFERENCES

Aoki, Y., and Muraoka, N. 1996. Optimizing plans and obtaining know-how for design with genetic algorithm. *Journal of Architecture, Planning and Environmental Engineering (Transactions of Architectural Institute of Japan)* 484:129–135 (in Japanese).

Cooper, L. 1963. Location-allocation problems. *Operations Research* 11:331-343.

Eldredge, N. 1995. *Reinventing Darwin: The great debate at the high table of evolutionary theory*. Hoboken, NJ: Wiley.

Eldredge, N., and Gould, S. J. 1972. Punctuated equilibria: an alternative to phyletic gradualism. In *Models in paleobiology*, ed. T. J. M. Schopf, 82–115. San Francisco: Freeman Cooper.

Hidaka, Y., and Asami, Y. 1998. A system for optimal division of lots based on genetic algorithm. *Papers and Proceedings of the Geographical Information Systems Association of Japan* 7:275–80 (in Japanese).

Holland, J. H. 1992. *Adaptation in natural and artificial systems*. 2nd ed. Cambridge, MA: MIT Press.

Hori, A., and Yoshikawa, T. 1999. A study on location-allocation planning of day facilities for the elderly considering capacity of facilities. *Summaries of Technical Papers of the Annual Meeting of Architectural Institute of Japan* F-1:669–70 (in Japanese).

Hori, A., and Yoshikawa, T. 2000. A new method of facility location by a genetic algorithm based on co-evolution. *Summaries of Technical Papers of the Annual Meeting of Architectural Institute of Japan* F-1:777–78 (in Japanese).

Hori, A., and Yoshikawa, T. 2001. A new method of facility location by a genetic algorithm based on co-evolution—Locational optimization of facilities by co-evolution of their locations and user allocation. *Journal of Architecture, Planning and Environmental Engineering (Transactions of Architectural Institute of Japan)* 540:221–27 (in Japanese).

Kauffman, S. 1995. *At home in the universe: The search for laws of self-organization and complexity*. Oxford: Oxford University Press.

Kawamura, H. 2000. Applicability of intelligent life systems to architectural structure systems. *Journal of Architecture and Building Science, Architectural Institute of Japan* 1450:36–37 (in Japanese).

Kawata, M. 1989. *A view of evolution theory*. Tokyo: Kinokuniya Shoten (in Japanese).

Kishimoto, T. 2000. Optimal location of facilities complexes with multi-layered structure. *Journal of Architecture, Planning and Environmental Engineering (Transactions of Architectural Institute of Japan)* 529:233–39 (in Japanese).

Koshizuka, T., and Kobayashi, J. 1983. On the relation between road distance and Euclidean distance. *Papers of the Annual Conference of the City Planning Institute of Japan* 18:43–48 (in Japanese).

Kurita, O. 1993. Approximate formulae for inter-regional distances based on the deterrent functions. *Papers on City Planning (Transaction of City Planning Institute of Japan)* 28:391–96 (in Japanese).

Minemasa, K., Ito, K., and Furusaka S. 1998. Development of strategic decision-making system of building detail design and exploring a scheduling system in genetic algorithm. *Journal of Architecture, Planning and Environmental Engineering (Transactions of Architectural Institute of Japan)* 512:229–36 (in Japanese).

Nosaka, M., and Yoshikawa, T. 1999. A study on location-allocation planning of day facilities for the elderly: A case study on estimated future population of Tama New Town. *Journal of Architecture, Planning and Environmental Engineering (Transactions of Architectural Institute of Japan)* 525:201–8 (in Japanese).

Ohsaki, J. 1998. Facility layout optimization based on Markov chain model and genetic algorithms. *Journal of Architecture, Planning and Environmental Engineering (Transactions of Architectural Institute of Japan)* 510:251–58 (in Japanese).

Okabe, A., and Suzuki, A. 1992. *Mathematics of optimal location*. Tokyo: Asakura Shoten (in Japanese).

Okunuki, K., and Okabe, A. 1995. Optimization of successively inclusive hierarchical facilities on a plane. *Papers on City Planning (Transactions of City Planning Institute of Japan)* 30:565–70 (in Japanese).

Osawa, Y. 1992. Facility location theory model. In *Analytic models for architecture and urban planning*, ed. Architectural Institute of Japan, 136–49. Tokyo: Inoue Shoin (in Japanese).

Ridley, M. 1993. *The red queen: Sex and the evolution of human nature*. New York: Viking Press.

Suzuki, T. 1990. A note on the optimal hierarchy system of the facilities. *Papers on City Planning (Transactions of City Planning Institute of Japan)* 25:331–36 (in Japanese).

Tagashira, N., and Okabe, A. 1998. Effect of spatial data units on the variance of a regression model containing a distance variable between a pre-determined point and the spatial data unit. *Theory and Applications of GIS* 6:29–38 (in Japanese).

Takabayashi, J. 1995. *Mystery of co-evolution: Ecology from the chemical viewpoint*. Tokyo: Heibonsha (in Japanese).

Takeda, Y. 1999. A location-allocation model with a space-time prism concept for an intermediate facility. *Geographical Review of Japan* 72A-11:721–45 (in Japanese).

Takizawa, A., Kawamura, H., and Tani A. 1998. Formation of urban land use pattern by genetic algorithm. *Journal of Architecture, Planning and Environmental Engineering (Transactions of Architectural Institute of Japan)* 495:281–87 (in Japanese).

Tamura, T., Masuya, Y., and Saito, K. 1994. Application of genetic algorithms to optimum location pattern of parking lots. *Papers on City Planning (Transaction of City Planning Institute of Japan)* 29:307–12 (in Japanese).

Tanimura, H. 1984a. Concept of local facility planning. In *Regional facility planning*, ed. Y. Uchida et al., 3–26. New Compendium Series on Architecture, vol. 21. Tokyo: Shokokusha (in Japanese).

Tanimura, H. 1984b. Local facility planning model. In *Regional facility planning*, ed. Y. Uchida et al., 309–86. New Compendium Series on Architecture, vol. 21. Tokyo: Shokokusha (in Japanese).

Tanimura, H. 1986. Mathematics of facility location planning. In *Mathematics of city planning*, ed. H. Tanimura, 56–96. Tokyo: Asakura Shoten (in Japanese).

Yamashita, G., and Tomokiyo, T. 1998. A tentative method to optimize the welfare service in the home for the elderly by genetic algorithms. *Journal of Architecture, Planning and Environmental Engineering (Transactions of Architectural Institute of Japan)* 509:105–12 (in Japanese).

13 Hierarchy of System of Urban Facilities

*Focusing on the Relationship between Administrative Systems and Population Agglomerations**

Kei-ichi Okunuki

13.1 INTRODUCTION

In this chapter, we will discuss how urban hierarchical systems and the geographical distribution of populations are related. Urban hierarchical systems are found in a variety of places, including those related to medicine, commerce, and transportation services. Here we will focus on hierarchical systems found within public services. The geographical divisions forming the jurisdictions of such public services as police and fire services can be taken as governmental geographical divisions most appropriate to the various types of public service. In actuality, however, a number of types of public service are often supplied as a group for a common governmental geographical division. That is not to say that all of the numerous public services are supplied within a single governmental region (for example, a country), as generally these services are supplied as a hierarchically divided system. Japan, for example, implements a hierarchical system that provides services at the national, prefectural, and municipal levels. The type of hierarchical system implemented for administration should be greatly influenced by the cities in which those persons engaged in public services live. Assuming a mono-level hierarchical system in which the national level is the only geographical division, then all public services would be supplied by the country and all engaged persons would be agglomerated in the city in which administrative facilities are located (in other words, the nation's capital). On the other hand, if governmental geographical divisions most appropriate to each individual public service are to be

* This chapter is revised and extended from Okunuki, K., "A Note on the Relationship between the Administrative Hierarchical Systems and the Number of Employees," *Papers on City Planning* 26 (1991): 553–58.

established, a hierarchical system with an extreme number of levels would result, and one would expect the persons engaged in those services to be scattered among multiple cities. We will make a theoretical examination of this using a simple model. In the literature, many researchers have investigated mathematical solutions to the problem of distribution of hierarchical facilities, the results of which are summarized by Sahin and Süral (2007). We will, however, avoid proposing new solutions for mathematical optimization and instead consider the geographical distribution of population, assuming optimal locations of facilities. We will first postulate that the number of persons engaged in the supply of a single type of public service is given by some function of the number of facilities related to that service. Next, we will calculate the number of persons engaged in that service for each city in which service facilities are located. By performing this calculation for a number of hierarchical systems, we will examine the relationship between the administration of urban hierarchical systems and the degree of agglomeration of population in cities.

13.2 A MODEL FOR OPTIMIZATION OF ONE-DIMENSIONAL DISTRIBUTION OF EMPLOYEES ENGAGED IN PUBLIC SERVICES

Here, we construct a mathematical model for the consideration of the distribution of employees engaged in public services. Within a subject region A the number of service levels is given by s, with fewer levels indicating higher-order services. The number of administrative facilities providing a level j service is given by n_j, with fewer facilities indicating higher-order services. Within those administrative facilities providing a level j service, we take the number of persons engaged in the provision of that service within a given facility as p_j. Letting P be the number of persons engaged in overall administration, we have

$$P = \sum_{j=1}^{s} n_j p_j \tag{13.1}$$

In the real world, we note that there are few cities with large populations, and numerous cities with small populations. Listing cities by the size of their population would result in nearly identical city sizes as the rank decreases. Zipf (1941) focused on this, and experimentally verified the existence of a rule relating to the population of a city and its rank in such a listing. The rule is extremely simple, stating that multiplying a city's population by its rank results in a constant. This rule is generally called the rank-size rule. For example, from the 2006 populations of U.S. cities (Figure 13.1), it would appear that the rank-size rule approximately holds. Taking hierarchical systems of urban facilities, too, it appears that there are few facilities that accommodate a large number of employees, and a large number of small facilities with a small number of employees. We shall therefore postulate that a service with a small number of facilities n_j will have a large number of engaged persons per facility p_j. We define services with a small number of facilities n_j as higher-order

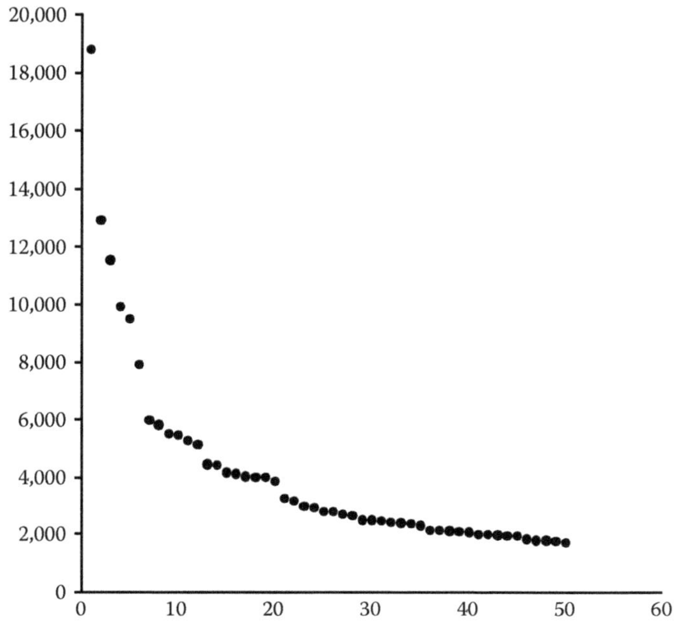

FIGURE 13.1 Rank and population of U.S. cities. The unit for the vertical axis is thousands; the unit for the horizontal axis is rank. Data obtained from Annual Estimates of the Population of Metropolitan and Micropolitan Statistical Areas: April 1, 2000 to July 1, 2006.

services, and so the higher the order of the service, the larger the number of engaged persons per facility p_j will be.

Next, we rank the facilities of all administrative systems in descending order of the number of persons they accommodate. Because there are n_j facilities supplying a public service of level j, the rank will be from

$$1+\sum_{k=1}^{j-1} n_k \tag{13.2}$$

to

$$\sum_{k=1}^{j} n_k \tag{13.3}$$

The total number of persons accommodated by a facility providing a level j public service is p_j. Let us take the ranking of a facility providing a level j public service to be the average of its ranking from Equation 13.2 to that of Equation 13.3. In other words, the rank of all facilities providing a level j service with p_j associated persons is given as

$$\frac{n_j+1}{2}+\sum_{k=1}^{j-1} n_k \tag{13.4}$$

Assuming that the rank-size rule holds for facility systems, we have

$$p_j \; \frac{n_j+1}{2} + \sum_{k=1}^{j-1} n_k \; = K \tag{13.5}$$

K, however, is a constant. From Equations 13.1 and 13.5, therefore, p_j is given by the following.

Number of employees associated with a level j service facility:

$$p_j = \frac{P}{\frac{n_j+1}{2} + \sum_{k=1}^{j-1} n_k \; \sum_{i=1}^{s} \frac{n_i}{\frac{n_i+1}{2} + \sum_{k=1}^{i-1} n_k}} \tag{13.6}$$

For example, if the number of facilities n_j is j (in other words, if there are j facilities for a level j service), the above equation is simplified even further, becoming

$$p_j = \frac{P}{\left(j^2+1\right)\sum_{i=1}^{s} \frac{i}{i^2+1}} \tag{13.7}$$

Let us next consider the location of facilities. We will take the minimization of the summation of the travel distance between all users and facilities as a minisum problem. Actual cities are spread out in two dimensions, and so it would be best to create a two-dimensional model. Our primary goal of finding a qualitative relationship between hierarchical structures and population agglomerations is simplified, however, by the use of a one-dimensional model for analysis. In the following discussion, therefore, we will consider a minisum problem on a one-dimensional area of length 1.

Let us first describe some assumptions related to facility usage. Let us take an even distribution of users in a finite area A as having a density of 1. We shall assume that when a user wishes to receive a level j service, that user will utilize the closest facility supplying a level j service, and that distances are calculated according to Euclidian (straight-line) distances. Numbering facilities that provide level j service from 1 to n_j, the range of persons utilizing the kth facility ($k \le n_j$) forms a Voronoi region ($V_k[j]$) in a Voronoi diagram generated by facilities 1 to n_j. In other words, taking the location vector of the kth facility as x_k, and x as an arbitrary location, $V_k[j]$ is given by

$$V_k^{[j]} = \left\{ x \,\middle|\, \|x - x_k\| \le \|x - x_h\|, h \ne k, k = 1,\ldots,n_j \right\} \tag{13.8}$$

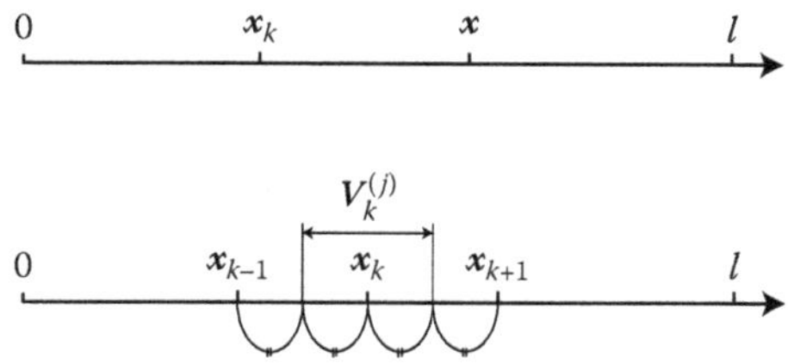

FIGURE 13.2 One-dimensional area (upper graph) and a Voronoi region for the *k*th level *j* service (lower graph).

Here, because the target area A is a one-dimensional area, as indicated in Figure 13.2, if we take the location vector of the kth facility as x_k, and x as an arbitrary point, then assume

$$0 < x_1 < \ldots < x_k < x_{k+1} < \ldots < x_{n_j} < l \tag{13.9}$$

Equation 13.8 becomes

$$V_1^{[j]} = \left. x \right| 0 \le x \le \frac{x_1 + x_2}{2} \tag{13.10}$$

$$V_k^{[j]} = \left. x \right| \frac{x_{k-1} + x_k}{2} \le x \le \frac{x_k + x_{k+1}}{2}, \; k = 2,\ldots,n_j - 1 \tag{13.11}$$

$$V_{n_s}^{[j]} = \left. x \right| \frac{x_{n_s-1} + x_{n_s}}{2} \le x \le l \tag{13.12}$$

The total travel distance T_j of users of a level j service is therefore

$$T_j = \sum_{k=1}^{n_j} \int_{V_k^{[j]}} \left\| x - x_k \right\| \mathrm{d}x \tag{13.13}$$

The locations of a level j service can be found by solving

to get the following:

$$\min_{x_1, \supset, x_{ns}} T_j \tag{13.14}$$

Locations of a level j service:

$$x_k = \frac{l}{n_j} \; k = \frac{1}{2}, \quad k = 1,\ldots,n_s \tag{13.15}$$

The above determine the number of employees associated with each service facility p_j, as well as facility locations x_k ($k = 1, \ldots, n_j$). Distribution of personnel throughout the entire administrative system involves the locations of p_j personnel in locations x_k ($k = 1, \ldots, n_j$) for each system, superimposing the personnel of all systems.

13.3 DISTRIBUTION OF PERSONNEL IN ADMINISTRATIVE SYSTEMS WITH A SUCCESSIVELY INCLUSIVE HIERARCHY

For the geographical division of public services, as previously discussed there should be a different optimal geographical division (Equations 13.10 to 13.12) for each service. In real-world administrative systems, however, services are divided and supplied from at least three levels: national, state or prefectural, and municipal. In most cases, such administrative systems have what Okabe et al. (1998) call a "successively inclusive hierarchy." A successively inclusive hierarchy is one in which the facilities for a given hierarchical level provide all of the services for that level and those below it. In Tokyo, for example, national-level organizations and metropolitan-level organizations share the same city. If metropolitan Tokyo is taken to be a single facility, then Tokyo contains not only national-level services, but also metropolitan-level and town-level services.

For administrative systems with a successively inclusive hierarchy, the number of facilities is determined according to the following principles (the central place theory of Christaller (1933) and Lösch (1954) indicates that the number of cities will depend on a marketing principle, a transport principle, and an administrative principle). First, the facility providing the highest-order service is placed. In the previous one-dimensional model, the highest-order service would be located at the center point of the region. This facility only will not handle lower-order services (Christaller (1933) explains this using the concept of travel costs). The number of facilities is then increased. Should the number of facilities be increased to two, then their locations according to Equation 13.15 would become

$$x_1 = \frac{l}{4}, \quad x_2 = \frac{3l}{4} \tag{13.16}$$

Because the administrative system has a successively inclusive hierarchy, however, it is necessary that the service be supplied from the facility already existing at the central point of the region. Equation 13.16 does not fulfill this requirement, and so we increase the number of facilities to three. The location according to Equation 13.15 then becomes

$$x_1 = \frac{l}{6}, \quad x_2 = \frac{l}{2}, \quad x_2 = \frac{5l}{6} \tag{13.17}$$

and this location meets the requirements of a successively inclusive hierarchy. By repeating this process, we form the hierarchical structure of the administrative systems. The principle for the formation of administrative systems is therefore that the ratio between the number of facilities at a given level and the number of facilities in the level immediately below it is 1:3. It therefore follows that in the case of an administrative system in one dimension, the number of administrative bodies would always be some multiple of 3. Considering location according to this principle in two dimensions, we see that the known ratio is 1:7 (Christaller, 1933), and the number of administrative bodies should therefore be some multiple of 7. In Japan, the actual number of administrative bodies in the administrative system is 47 at the prefectural level and 3,245 at the municipal level (as of 1990), numbers very close to the second and fourth powers of 7, and so lending support to the principle described above. In the new federal systems that are considered in Japan from time to time, the number of federal districts proposed is in the order of the first power of 7, and the number of cities in municipal association plans such as the wide-area municipal association is approximately the third power of 7. Consideration of systems of federation or municipality is due to the fact that current administrative systems have reached hierarchical constructions nearing the second and fourth powers of 7, and the creation of administrative systems at the levels of the first or third power of 7 would increase the efficiency of public services.

When current administrative systems (in other words, the hierarchical structure comprised of the three levels of national, prefectural, and municipal services) are formed upon a one-dimensional curve, the number of facilities for each service level becomes one at the national level, 9 (the second power of 3) at the prefectural level, and 81 (the fourth power of 3) at the municipal level. Let us use a one-dimensional model to examine how to distribute personnel in such an administrative system.

In Equation 13.6, the number of persons p_j associated with a level j service was obtained as a function of the number of facilities n_j. In the administrative system that we are considering there are three service levels, and the number of facilities for each level is

$$n_1 = 1, \quad n_2 = 9, \quad n_2 = 81 \tag{13.18}$$

If we insert these values into Equation 13.6, the numbers of employees associated with each facility p_j become

$$p_1 = \frac{P}{3}, \quad p_2 = \frac{P}{27}, \quad p_2 = \frac{P}{243} \tag{13.19}$$

Because Equation 13.15 gives locations for facilities at each service level, it is possible to see how the personnel for each service level must be distributed. Overlaying this, the distribution of personnel for the administrative systems will be as shown in Figure 13.3. From that figure, we can see that personnel are collected near the central point (37% of all personnel are agglomerated there). In real-world administrative systems, too, we see agglomerations of populations at capitals, lending support to the ideas presented here.

Next, let us consider how personnel might be distributed should current administrative systems (national, prefectural, and municipal) be modified to a system of 1 national, 3 (first power of 3) federal, and 27 (third power of 3) municipal unions. The "doshu" system is one example of a proposed regional system reform plan compatible

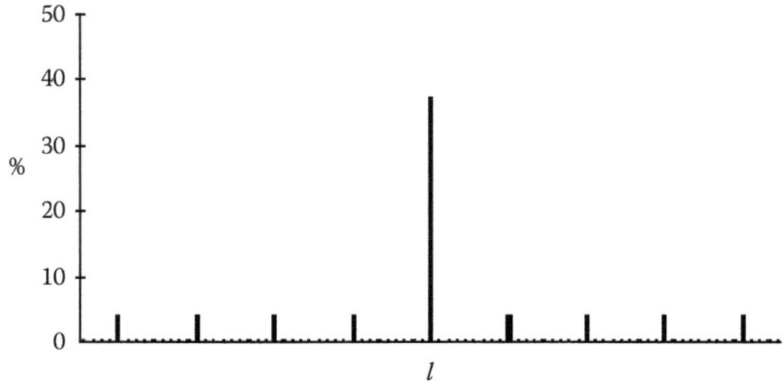

FIGURE 13.3 Distribution of employees in prefectural and municipal systems; vertical axis indicates employees in city as a percentage of the overall total.

with such an administrative system. Let us examine it using a one-dimensional model.

In Equation 13.6, the number of persons p_j associated with a level j service was obtained as a function of the number of facilities n_j. In the administrative system that we are considering there are three service levels, and the number of facilities for each level is

$$n_1 = 1, \quad n_2 = 3, \quad n_3 = 27 \tag{13.20}$$

If we insert these values into Equation 13.6, the numbers of employees associated with each facility p_j become

$$p_1 = \frac{P}{3}, \quad p_2 = \frac{P}{9}, \quad p_3 = \frac{P}{81} \tag{13.21}$$

Because Equation 13.15 gives locations for facilities at each service level, it is possible to see how the personnel for each service level must be distributed. Overlaying this, the distribution of personnel for the administrative systems will be as shown in Figure 13.4. From that figure, we can see that personnel are collected near the central point (46% of all personnel are agglomerated there). This indicates a possibility that administrative systems formed from national, federated state, and municipal association systems may further the agglomeration of the population in the capital. On the other hand, however, two other cities of larger-than-before population are also formed. It therefore seems that this administrative system is a transition from a unipolar population agglomeration to a multipolar one.

According to our examination above, we have seen that while the introduction of a federal or municipal association system will promote a shift from a unipolar to a multipolar construction, it also poses the risk of further population agglomeration within the capital. What, then, might we do to lower the agglomeration in the capital to a level below that which is currently seen? The agglomeration of personnel at the center point of our one-dimensional model occurs because all hierarchies provide

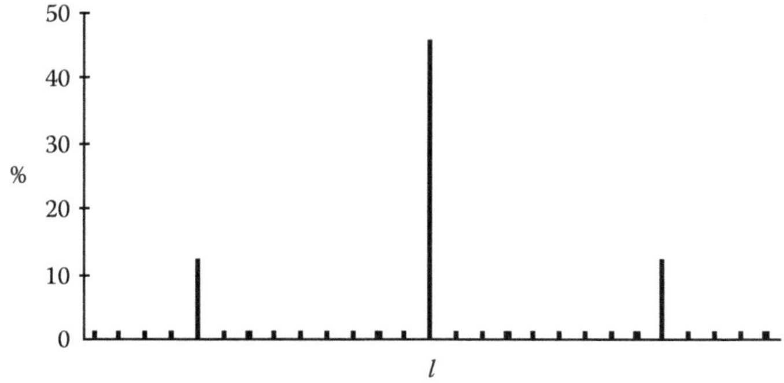

FIGURE 13.4 Employee distribution under a federal/municipal association system; vertical axis indicates employees in city as a percentage of the overall total.

their services at that point. Does it not follow, therefore, that were the hierarchical system to be such that construction was from levels in which the number of associated persons per unit facility was lowest, then the agglomeration at the center point could be lessened? Let us confirm this idea.

Beyond the country level, the two levels to be chosen as having the smallest number of associated persons per unit facility are the associated city level and the municipal level. Let us therefore consider a one-dimensional model of the distribution of employees associated with administrative systems composed of national, associated city, and municipal levels.

In Equation 13.6, the number of persons p_j associated with a level j service was obtained as a function of the number of facilities n_j. In the administrative system that we are considering there are three service levels, and the number of facilities for each level is

$$n_1 = 1, \quad n_2 = 27, \quad n_3 = 81 \tag{13.22}$$

If we insert these values into Equation 13.6, the numbers of employees associated with each facility p_j become

$$p_1 = \frac{P}{3}, \quad p_2 = \frac{P}{81}, \quad p_3 = \frac{P}{243} \tag{13.23}$$

Because Equation 13.15 gives locations for facilities at each service level, it is possible to see how the personnel for each service level must be distributed. Overlaying this, the distribution of personnel for the administrative systems will be as shown in Figure 13.5. From that figure, we can see that personnel are collected near the central point (35% of all personnel are agglomerated there). While this shows some weakening, it does not indicate the potential for significant alleviation of agglomeration. The effect of this administrative system, in fact, is a promotion of distribution. In other words, compared with the two previously described administrative systems, personnel are further distributed. While overagglomeration of the system is not solved, it is possible at least to halt depopulation.

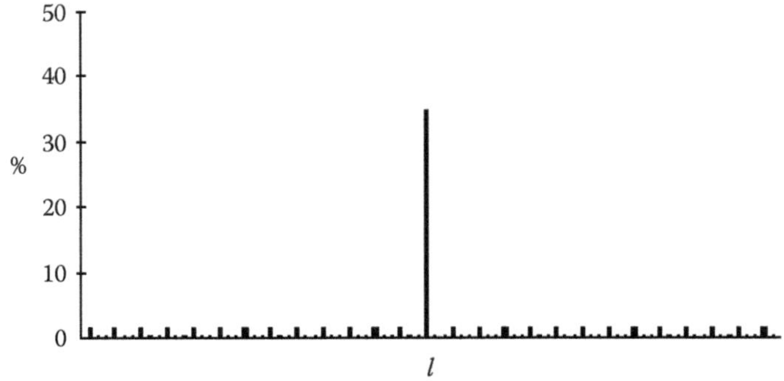

FIGURE 13.5 Employee distribution under an associated city/metropolitan system; vertical axis indicates employees in city as a percentage of the overall total.

Up to this point we have considered three administrative systems, but have seen that none of them really promise significant improvements in the current overagglomeration. This is because of the small number of administrative system levels and the fact that the administrative systems have successively inclusive hierarchies. As long as successively inclusive hierarchies are to be retained, the only way to improve excessive agglomeration is to increase the number of levels. We would therefore like to consider a one-dimensional model of personnel distribution for an administrative system composed of five levels: national, federal, prefectural, associated city, and municipal.

In Equation 13.6, the number of persons p_j associated with a level j service was obtained as a function of the number of facilities n_j. In the administrative system that we are considering there are five service levels, and the number of facilities for each level is

$$n_1 = 1, \quad n_2 = 3, \quad n_3 = 9, \quad n_4 = 27, \quad n_5 = 81 \tag{13.24}$$

If we insert these values into Equation 13.6, the numbers of employees associated with each facility p_j become

$$p_1 = \frac{P}{5}, \quad p_2 = \frac{P}{15}, \quad p_3 = \frac{P}{45}, \quad p_4 = \frac{P}{135}, \quad p_5 = \frac{P}{405} \tag{13.25}$$

Because Equation 13.15 gives locations for facilities at each service level, it is possible to see how the personnel for each service level must be distributed. Overlaying this, the distribution of personnel for the administrative systems will be as shown in Figure 13.6. From that figure, we can see that the collection of personnel near the central point is significantly reduced (30% of all personnel are agglomerated there).

Our investigation of one-dimensional models up to this point has shown that current administrative systems (national, prefectural, and municipal) make the development of overagglomeration more likely than does other systems, that federal and

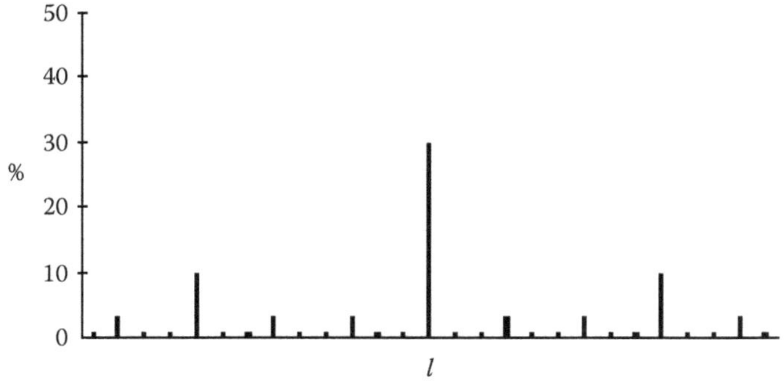

FIGURE 13.6 Employee distribution under a federal/prefectural/associated city/metropolitan system; vertical axis indicates employees in city as a percentage of the overall total.

associated city systems promote multipolar agglomerations, that associated city and municipal systems promote dispersal, and that an administrative system with an increased number of levels to national, federal, prefectural, associated city, and municipal levels may lead to a lessening of overagglomeration. Given an administrative system with a successively inclusive hierarchy, the only way to lessen the agglomeration of distribution at one pole is to increase the number of levels, and administrative hierarchical structures with lower levels are effective in the promotion of distribution. Furthermore, multipolar agglomerations can be promoted by administrative hierarchical structures that have higher levels.

13.4 CONCLUSION

In this investigation, we have mathematically analyzed the relationship between urban hierarchical structures and urban agglomeration. The one-dimensional models used here are simple, but are sufficient for analysis of the qualitative aspects of hierarchical constructs and agglomeration, and their simplicity aids ease of understanding. For administrative systems with a successively inclusive hierarchy, we have seen that current administrative systems (national, prefectural, and municipal) make it easy for excessive agglomeration to arise. Our investigation suggests that the current federal and associated city plans being considered will not contribute much to a reduction in excess agglomeration, though they will promote a shift to multipolarity in agglomeration. Furthermore, associated city and municipal systems are effective in the promotion of dispersion, and an increase in system levels would be effective in lessening the degree of excess agglomeration.

REFERENCES

Christaller, W. 1933. *Die zentralen orte in Süddeutschland*. Jena, Germany: Fisher. English translation by Carlisle Baskin, W. 1966. *The central places of Southern Germany*. Englewood Cliffs, NJ: Prentice Hall.

Lösch, A. 1954. *The economics of location*. New Haven, CT: Yale University Press.

Okabe, A., Okunuki, K., and Suzuki, T. 1998. A computational method for optimizing the hierarchy and spatial configuration of successively inclusive facilities on a continuous plane. *Location Science* 5:255–68.

Şahin, G., and Süral, H. 2007. A review of hierarchical facility location models. *Computers & Operations Research* 34:2310–31.

Zipf, G.K. 1941. *National unity and disunity*. Bloomington, IN: Principia Press.

Author Index

Subject Index

U

V

W

For Product Safety Concerns and Information please contact our EU representative GPSR@taylorandfrancis.com
Taylor & Francis Verlag GmbH, Kaufingerstraße 24, 80331 München, Germany

www.ingramcontent.com/pod-product-compliance
Ingram Content Group UK Ltd.
Pitfield, Milton Keynes, MK11 3LW, UK
UKHW021722240926
475702UK00029BA/1840
9780367385484